AF182526

NMR 33

Basic Principles and Progress

Editors:
P. Diehl E. Fluck H. Günther R. Kosfeld J. Seelig

Guest-Editor: B. Blümich

Solid-State NMR IV
Methods and Applications of Solid-State NMR

Guest-Editor: B. Blümich

With contributions by
A. E. Bennett, B. F. Chmelka, H. Eckert,
R. G. Griffin, S. Vega, J. W. Zwanziger

With 112 Figures and 1 Table

Springer-Verlag
Berlin Heidelberg NewYork
London Paris Tokyo
Hong Kong Barcelona Budapest

ISBN-13:978-3-642-79129-1 e-ISBN-13:978-3-642-79127-7
DOI: 10.1007/978-3-642-79127-7

Softcover reprint of the hardcover 1st edition 1994
Library of Congress Catalog Card Number 93-9522

Typesetting: Thomson Press (India) Ltd, New Delhi

SPIN: 10128818 51/3020 - 5 4 3 2 1 0 - Printed on acid-free paper

Preface

Solid-State NMR is a branch of Nuclear Magnetic Resonance which is presently experiencing a phase of strongly increasing popularity. The most striking evidence is the large number of contributions from Solid-State Resonance at NMR meetings, approaching that of liquid-state resonance. Important progress can be observed in the areas of methodological developments and applications to organic and inorganic matter. One volume devoted to more or less one of each of these areas has been published in the preceding three issues. This volume can be considered an addendum to this series.

Selected methods and applications of Solid-State NMR are featured in three chapters. The first one treats the recoupling of dipolar interactions in solids, which are averaged by fast sample rotation. Following an introduction to effective Hamiltonians and Floquet theory, different types of experiment such as rotary resonance, dipolar chemical shift correlation spectroscopy, rotational resonance and multipulse recoupling are treated in the powerful Floquet formalism. In the second chapter, the different approaches to line narrowing of quadrupolar nuclei are reviewed in a consistent formulation of double resonance (DOR) and dynamic angle spinning (DAS). Practical aspects of probe design are considered as well as advanced 2D experiments, sensitivity enhancement techniques, and spinning sideband manipulations. The use of such techniques dramatically increases the number of nuclei which can be probed in high resolution NMR spectroscopy. The final chapter describes new experimental approaches and results of structural studies of noncrystalline solids. A wide variety of solid-state techniques for investigations of glasses is reviewed. Examples are wideline NMR, MAS NMR, and 2D NMR including multiple quantum methods. Application of these techniques to systematic studies of different glasses are presented in the main part, illustrating the new frontiers in understanding the complexity of glasses which can be explored by the use of modern solid-state NMR methods.

The authors of this volume are thanked in particular for their dedication in writing the contributions. Springer-Verlag has been very helpful in its assistance and editorial supervision.

Aachen, May 1994

B. Blümich
R. Kosfeld

Guest-Editor

Prof. Dr. Bernhard Blümich
Lehrstuhl für Makromolekulare Chemie,
RTWH Aachen, Worringer Weg 1,
D-52056 Aachen, FRG

Table of Contents

Tables of Contents to Volumes 30 and 31

Solid-State NMR I - Methods

Solid-State NMR II - Inorganic Matter

Table of Contents to Volume 32

Solid State NMR III - Organic Matter

Recoupling of Homo- and Heteronuclear Dipolar Interactions in Rotating Solids

Andrew E. Bennett[1], Robert G. Griffin[1], and Shimon Vega[2]

[1]Francis Bitter National Magnet Laboratory and Department of Chemistry, Massachusetts Institute of Technology, Cambridge, MA 02139, USA
[2]Department of Chemical Physics, Weizmann Institute of Science, Rehovot 76100, Israel

Table of Contents

NMR Basic Principles and Progress, Vol. 33

In recent years, several new techniques have been developed to measure dipolar couplings in powdered solids using magic angle spinning (MAS) NMR spectroscopy. These experiments have made it possible to measure interatomic distances in polycrystalline and amorphous solids, providing information about their molecular structure that is often difficult or impossible to obtain by other methods. All of these approaches take advantage of the improved resolution and sensitivity of "dilute" spins under MAS, and where appropriate, cross polarization and proton decoupling. In addition, they involve the application of some means of restoring selected dipolar couplings to the MAS experiment so that internuclear distances may be measured. In this article, a variety of homo- and heteronuclear recoupling experiments are reviewed and analyzed within the frameworks of Average Hamiltonian Theory (AHT) and Floquet Theory. Selected applications illustrating the utility of these techniques are also discussed.

1 Introduction

The measurement of homo- and heteronuclear dipolar couplings by nuclear magnetic resonance (NMR) techniques is an important tool for the determination of molecular structure in solids. In a static polycrystalline solid, the dipolar coupling between two magnetically dilute spins results in the characteristic "Pake pattern" [1], first observed in the ^1H spectrum of gypsum, $CaSO_4 \cdot 2H_2O$, which arises from the interaction between the two protons in the water molecules of hydration. The splitting between the singularities provides a straightforward measurement of the dipolar coupling constant and therefore the internuclear distance between the two spins. Unfortunately, in the more general case, the structural information revealed by internuclear distances cannot be obtained directly from the static ^1H NMR spectrum because of the multiplicity of couplings. In situations involving other nuclei, such as ^{13}C, ^{15}N, and ^{31}P, large chemical shift anisotropies, as well as other line broadening mechanisms, obscure the lineshape perturbations from the through-space dipolar couplings.

As a consequence of these problems, various methods have been developed that separate the dipolar couplings of interest from the other interactions in static samples, most notably the chemical shift terms present in the spin Hamiltonian. These techniques include Separated Local Field (SLF) spectroscopy [2–5], Spin Echo Double Resonance (SEDOR) [6–11], nutation NMR [12], and the use of Carr-Purcell echo sequences to measure homonuclear dipolar couplings [12–14]. The two-dimensional SLF experiment, performed on both single crystals [2] and powders [3], correlates the anisotropic chemical shift interactions of dilute spins with their dipolar couplings to neighboring abundant spins. In SEDOR, the intensity of a spin echo is attenuated by heteronuclear dipole-dipole couplings that are prevented from refocusing by the application of suitable π pulse sequences. By comparison with the amplitude of a complete spin echo, these couplings can be measured to determine internuclear distances. In rotating solids, a similar strategy is employed in Rotational Echo Double Resonance (REDOR) experiments [15, 16], where the formation of rotational echoes arising from the heteronuclear dipolar coupling is hindered by certain rotor-synchronized π pulse sequences. The remaining two experiments separate dipolar couplings from chemical shifts in static homonuclear spin systems. In nutation NMR, this separation is achieved by the application of continuous radio-frequency (RF) fields, which eliminate chemical shifts and heteronuclear interactions but retain scaled homonuclear couplings. Carr-Purcell echo sequences, on the other hand, retain the full strength of the homonuclear interaction in the limit of short π pulses.

These approaches are useful for the determination of molecular structure in static solids, including powders. However, in order to obtain high resolution NMR spectra of solids, the application of magic angle spinning (MAS) is essential [17–22]. The combined use of proton decoupling, cross polarization (CP) [23], and MAS, which is now known as the CP/MAS experiment [19], has become the standard method for observing the high resolution spectra of "dilute" spin-half nuclei, such as ^{13}C, ^{15}N, and ^{31}P. MAS efficiently attenuates weak heteronuclear and homonuclear dipole-dipole couplings, as well as all chemical shift anisotropies, resulting in well-resolved spectra of the isotropic chemical shifts. In order to observe selected dipolar couplings in rotating solids, it is therefore necessary to reintroduce them into the experiment by some method that reverses their attenuation by MAS. At the same time, as in the case of non-rotating solids, it is frequently desirable to separate recoupled interactions from other contributions to the spin Hamiltonian. Often, this criterion leads to the elimination of chemical shifts from the experiment, at least in the sense of coherent averaging, through the use of spin echo sequences.

Several approaches have now been developed that achieve a suitable recoupling of the dipolar interaction for both heteronuclear and homonuclear spins in rotating solids. In the absence of RF pulses, the heteronuclear spin system is completely refocused after each rotor cycle and is therefore "inhomogeneous" in the terminology introduced by Maricq and Waugh (MW) [22]. The fact that all terms in the MAS spin Hamiltonian commute with one another at different times, a property that also holds under π pulse sequences, has led to several elegant spin echo techniques designed to measure heteronuclear distances [15, 24–27]. In contrast, under most conditions, the flip-flop term of the homonuclear dipolar coupling results in the non-commutation of the spin Hamiltonian with itself at different times, rendering it an "homogeneous" interaction in the MW sense. In the special case of a homonuclear spin pair, the chemical shift terms and the flip-flop part of the dipole-dipole interaction are non-commuting and lead to homogeneous behavior [22]. More generally, in the case of several spins, there are many dipolar couplings, and in most cases these interactions fail to commute because of the flip-flop terms [28]. Accordingly, the rotational refocusing of observed signals is spoiled to varying degrees, and it is often necessary to spin much faster than the magnitude of the interactions in order to suppress them efficiently. In the case of homonuclear spins, the development of dipolar recoupling experiments requires consideration of the additional complexity introduced by the homogeneous nature of the spin Hamiltonian.

In this review, various methods of dipolar recoupling are examined within the framework of effective spin Hamiltonians. Average Hamiltonian Theory (AHT) [29, 30] and Floquet Theory [31–34] provide two approaches to the construction of effective Hamiltonians in coupled spin systems governed by periodic interactions and perturbations, which in the present context comprise the combined influences of MAS and rotor-synchronized multiple pulse sequences. The next section presents the Hamiltonian appropriate for dipole-coupled spin systems and introduces the AHT and Floquet approaches. In the third section, we discuss several methods that are applicable to heteronuclear spin systems, including Rotary Resonance Recoupling (R^3) [35–37], Dipolar Chemical Shift Correlation Spectroscopy [24, 26, 38], and Rotational Echo Double Resonance experiments [15, 16, 39]. The following section describes several methods that are applicable to homonuclear spin systems. These techniques include experiments in which recoupling is driven by the rotor, as in longitudinal exchange at rotational resonance (R^2) [40–44], and methods in which multiple pulse sequences are applied in order to reintroduce the dipolar coupling. These methods can be further divided into direct recoupling sequences [45, 46], in which a train of pulses similar to that of a solid echo sequence is used to interfere with dipolar cancellation by MAS, and spin echo sequences employing π pulses [47, 48], where the recoupling effect depends on the modulation of the dipolar coupling by chemical shift differences between the spins in the toggling frame. In all cases where multiple pulse sequences are applied to achieve dipolar recoupling, the dynamics of the spin system under both RF pulses and magic angle spinning must be considered simultaneously. Finally, in order to illustrate the practical potential of dipolar recoupling methods, we discuss two applications to the determination of protein structure in the solid state.

2 Theoretical Framework for Recoupling Experiments

2.1 *The MAS Hamiltonian for Coupled Spin Systems*

Our discussion is restricted to systems consisting of coupled nuclei with spin angular momentum $I = 1/2$ that are subject to mechanical rotation about a fixed axis with respect to the magnetic field. In addition, RF fields are often applied to the spin system, introducing a second time-dependence into the MAS experiment. The relevant spin Hamiltonian of a rotating system can be written as follows [30, 49]:

$$H(t) = H_D(t) + H_{CS}(t) + H_J + H_{RF}(t), \tag{1}$$

where $H_D(t)$ represents the dipolar couplings among the spins, $H_{CS}(t)$ the contribution from the chemical shift terms, H_J the indirect spin-spin interactions, or J-couplings, and $H_{RF}(t)$ the RF fields applied to the spin system. Under MAS, the geometric factors describing the anisotropic parts of the various spin interactions are time-dependent. Explicit expressions for the terms of the spin Hamiltonian are given by the following expressions applicable to a set of spins $i,j,\ldots$, residing in a large magnetic field and subject to a rotating frame transformation with respect to all spins involved [30, 49]:

$$H_D(t) = \sum_{i<j} \omega_{ij}^{D}(t)[3I_{zj}I_{zj} - \vec{I}_i \cdot \vec{I}_j]; \tag{2a}$$

$$H_{CS}(t) = -\sum_{i} \omega_i^{CS}(t) I_{zi}; \tag{2b}$$

$$H_J = \sum_{i<j} \omega_{ij}^J \vec{I}_i \cdot \vec{I}_j; \tag{2c}$$

$$H_{RF}(t) = \sum_i \{\omega_{xi}^{RF}(t) I_{xi} + \omega_{yi}^{RF}(t) I_{yi}\} \tag{2d}$$

In the case of the J-coupling, the time-dependent anisotropic terms are assumed to be small and are therefore neglected. The part of the dipolar Hamiltonian in Eq. (2a) involving angular momentum operators is $[3I_{zi}I_{zj} - \vec{I}_i \cdot \vec{I}_j]$ for homonuclear spin pairs, and for heteronuclear spin pairs, the expression is further truncated to the form [10, 50]:

$$H_D(t) = \sum_{i<j} 2\omega_{ij}^D(t) I_{zi} I_{zj} \tag{3}$$

The two cases are distinguished by the flop-flop operator $[I_{+i}I_{-j} + I_{-i}I_{+j}]$, which appears only in the homonuclear case. In what follows below, we use the expression $I_{zi}S_{zj}$ to refer to the heteronuclear coupling between two nuclei i and j in order to distinguish it from the homonuclear Hamiltonian.

The dipolar coefficients $\omega_{ij}^D(t)$ are equal to the nuclear dipole-dipole interaction strength ω_{ij}^D between the nuclei i and j, multiplied by a geometric factor. In angular units, the dipolar coupling constant is given by:

$$\omega_{ij}^D = \frac{\mu_0}{4\pi} \frac{\gamma_i \gamma_j \hbar}{r_{ij}^3}, \tag{4}$$

where $\vec{r}_{ij}$ is the vector connecting the two interacting nuclei, and the factors γ_i and γ_j are the gyromagnetic constants of the spins. Knowledge of the coupling constant defined by Eq. (4) implies knowledge of the internuclear distance. In a rotating sample, the time-dependence of the dipolar interaction coefficient can be expressed in the form [51]:

$$\omega_{ij}^D(t) = \omega_{ij}^D \{G_0 + G_1 \cos(\omega_R t + \phi_{ij}) + G_2 \cos(2\omega_R t + 2\phi_{ij})\}, \tag{5}$$

where:

$$\begin{aligned} G_0 &= -\frac{(3\cos^2\theta_m - 1)}{2}\frac{(3\cos^2\theta_{ij} - 1)}{2}; \\ G_1 &= \tfrac{3}{4}\sin 2\theta_m \sin 2\theta_{ij}; \\ G_2 &= -\tfrac{3}{4}\sin^2\theta_m \sin^2\theta_{ij} \end{aligned} \tag{6}$$

The angle θ_m defines the relationship between the external magnetic field and the rotor axis, while (θ_{ij}, ϕ_{ij}) are the polar angles of $\vec{r}_{ij}$ in a rotating reference frame defined by the sample rotor at time $t = 0$. Figure 1 provides an illustration of the relationships among the laboratory frame, the rotor frame, and the internuclear vector. At the magic angle, $\theta_m = \cos^{-1}(1/\sqrt{3}) \approx 54.7°$, the contribution G_0 vanishes, and the MAS dipolar Hamiltonian contains no terms that are time-independent. The two remaining components in Eq. (5) are oscillatory and average to zero over each rotor period. Application of the formalism of

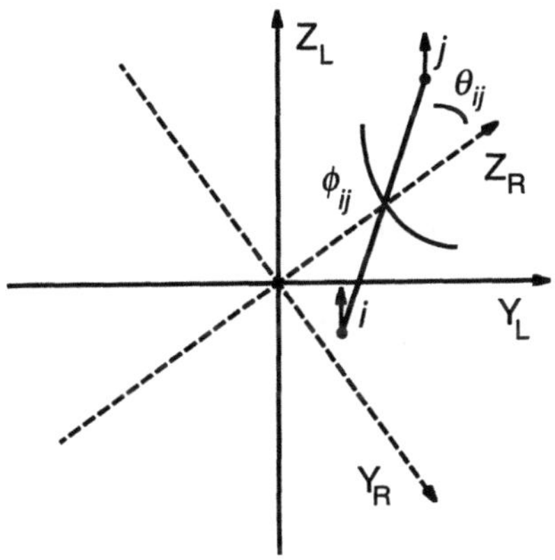

Fig. 1. Relationship of the rotor PAS to the laboratory reference frame. The axis of rotation Z_R is oriented at the magic angle θ_m with respect to the direction of the static magnetic field Z_L. The internuclear vector $\vec{r}_{ij}$ spins about the longitudinal axis of the rotor, and its time average therefore lies along the magic angle, where the dipolar coupling is zero. The X axes of the reference systems are parallel and point outward. While only two angles (θ_{ij}, ϕ_{ij}) are needed to express the orientation between the rotor and dipolar principal axis systems, three Euler angles must be specified for the CSA interaction

Floquet Theory requires the expansion of the MAS Hamiltonian in a Fourier series. The Fourier series of $\omega_{ij}^{D}(t)$ can be expressed as follows:

$$\omega_{ij}^{D}(t) = \sum_{n=-2}^{2} 2\omega_{n}^{ij} \exp\{in\omega_R t\}, \tag{7}$$

where:

$$\omega_{0}^{ij} = \tfrac{1}{2}\omega_{ij}^{D} G_0;$$
$$\omega_{\pm n}^{ij} = \tfrac{1}{4}\omega_{ij}^{D} G_{|n|} \exp\{\pm in\phi_{ij}\} \tag{8}$$

In a similar way, the chemical shift (CS) coefficient $\omega_{i}^{CS}(t)$ appearing in Eq. (2b) is a function of the three principal values $(\sigma_{11}^{i}, \sigma_{22}^{i}, \sigma_{33}^{i})$ of the chemical shift tensor in the principal axis system (PAS) of spin i, including the shift in frequency from resonance. The frequency offset, the chemical shift anisotropy (CSA), and the asymmetry factor are defined by:

$$\Delta\omega_i = \omega_0 \bar{\sigma}^i = \tfrac{1}{3}\omega_0(\sigma_{11}^{i} + \sigma_{22}^{i} + \sigma_{33}^{i});$$
$$\omega_{i}^{CSA} = \omega_0(\sigma_{33}^{i} - \bar{\sigma}^i);$$
$$\eta_i = \frac{(\sigma_{22}^{i} - \sigma_{11}^{i})}{(\sigma_{33}^{i} - \bar{\sigma}^i)}; \tag{9}$$

with the convention $|\sigma_{33}^{i} - \bar{\sigma}^i| \geqq |\sigma_{22}^{i} - \bar{\sigma}^i| \geqq |\sigma_{11}^{i} - \bar{\sigma}^i|$, where $\bar{\sigma}^i$ is the isotropic chemical shift of spin i. The Larmor frequency has the definition: $\omega_0 = -\gamma H_0$. In a way similar to the dipolar amplitude, the CS coefficient in a spinning solid has the general form [22, 51]:

$$\omega_{i}^{CS}(t) = \Delta\omega_i + \omega_{i}^{CSA}\{g_0 + g_1\cos(\omega_R t + \psi_1) + g_2\cos(2\omega_R t + \psi_2)\}, \tag{10}$$

where $(\alpha_i, \beta_i, \gamma_i)$ are the Euler angles of the CSA tensor of spin i in the reference frame defined by the rotor. The geometric factors are defined by the following expressions:

$$g_0 = \frac{(3\cos^2\theta_m - 1)}{2}\left\{\frac{(3\cos^2\beta_i - 1)}{2} - \frac{\eta_i}{2}\sin^2\beta_i \cos 2\gamma_i\right\};$$

$$g_1 = -\frac{1}{2}\sin 2\theta_m \sin\beta_i\{(\eta_i\cos 2\gamma_i + 3)^2\cos^2\beta_i + \eta_i^2\sin^2 2\gamma_i\}^{1/2};$$

$$g_2 = \frac{1}{2}\sin^2\theta_m\left\{\left[\frac{3}{2}\sin^2\beta_i - \frac{\eta_i}{2}\cos 2\gamma_i(1+\cos^2\beta_i)\right]^2 + \eta_i^2\cos^2\beta_i\sin^2 2\gamma_i\right\}^{1/2}; \tag{11}$$

and:

$$\psi_1 = \alpha_i + \tan^{-1}\left\{\frac{\eta_i \sin 2\gamma_i}{(\eta_i\cos 2\gamma_i + 3)\cos\beta_i}\right\};$$

$$\psi_2 = 2\alpha_i + \tan^{-1}\left\{\frac{-\eta_i\cos\beta_i\sin 2\gamma_i}{3/2\sin^2\beta_i - \eta_i/2\cos 2\gamma_i(1+\cos^2\beta_i)}\right\} \tag{12}$$

The time-independent term g_0 again vanishes at the magic angle. The Euler angle α_i appears only in the phase angles of rotation ψ_1 and ψ_2. We express the Fourier expansion of the coefficient $\omega_i^{CS}(t)$ as follows:

$$\omega_i^{CS}(t) = \sum_{n=-2}^{2} 2\omega_n^i \exp\{in\omega_R t\}, \tag{13}$$

using the definitions:

$$\omega_0^i = \tfrac{1}{2}\Delta\omega_i + \tfrac{1}{2}\omega_i^{CSA} g_0;$$

$$\omega_{\pm|n|}^i = \tfrac{1}{4}\omega_i^{CSA} g_{|n|}\exp\{\pm i\psi_{|n|}\} \tag{14}$$

In contrast to the dipolar coupling, all of whose components are oscillatory at the magic angle, the CS interaction tensor has, in general, a non-zero Fourier component at $n = 0$, namely, the resonance offset $\Delta\omega_i$. If we regard the interactions as second-rank tensors acting on the spin operators, this distinction arises from the fact that the trace of the matrix characterizing the dipolar coupling is zero, whereas the trace of the CS interaction is, in general, non-vanishing [30, 49].

Because we neglect the anisotropic part of the J-coupling Hamiltonian H_J, its coefficients are independent of orientation, and H_J remains independent of time during MAS experiments. In addition, the J-coupling of Eq. (2c) is truncated to the expression:

$$H_J = \omega_{ij}^J I_{zi} I_{zj}, \tag{15}$$

for heteronuclear spins, and also for homonuclear spins when $|\omega_i^{CS}(t) - \omega_j^{CS}(t)| \gg |\omega_{ij}^J|$ holds at all times. In most cases, we omit the J-coupling from our treatment of coupled spins because in solids it is much smaller than the direct dipole-dipole coupling, and consequently plays a subordinate role in nuclear spin dynamics.

The RF irradiation term $H_{RF}(t)$ of the Hamiltonian is in most cases strongly time-dependent. In many MAS experiments, the time-dependence of the RF field is periodic and synchronous with the rotation of the sample; therefore, transformation of the Hamiltonian into the toggling frame provides a means of simplifying the evaluation of spin evolution [30].

2.2 *Fictitious Spin-Half Operators*

Fictitious spin-half operators [52, 53] provide a convenient framework for expressing the various interactions present in spin systems and their commutation relations. Within a general manifold of spin states, the fictitious spin-half operators I_x^{pq}, I_y^{pq}, and I_z^{pq}, involving the two levels $|p\rangle$ and $|q\rangle$, with $p < q$, are characterized by the following non-zero matrix elements:

$$\langle p|I_x^{pq}|q\rangle = \langle q|I_x^{pq}|p\rangle = \frac{1}{2};$$

$$\langle p|I_y^{pq}|q\rangle = -\langle q|I_y^{pq}|p\rangle = -\frac{i}{2};$$

$$\langle p|I_z^{pq}|p\rangle = -\langle q|I_z^{pq}|q\rangle = \frac{1}{2} \qquad (16)$$

It is also convenient to define diagonal operators I_z^{pp} having the non-vanishing matrix element:

$$\langle p|I_z^{pp}|p\rangle = \tfrac{1}{2} \qquad (17)$$

These matrix elements define Hermitian operators. The fictitious spin-half operators are especially useful for the analysis of coherence and population between two levels within a larger system of states. The operators I_x^{pq}, I_y^{pq}, and I_z^{pq} obey the usual commutation relations of angular momentum:

$$[I_i^{pq}, I_j^{pq}] = iI_k^{pq}, \qquad (18)$$

where (i, j, k) are cyclic permutations of the directions (x, y, z) [52, 53].

In the particular case of two spin-half nuclei, the four possible spin states can be represented as follows:

$$|1\rangle = |\alpha\alpha\rangle; \quad |2\rangle = |\alpha\beta\rangle;$$

$$|3\rangle = |\beta\alpha\rangle; \quad |4\rangle = |\beta\beta\rangle \qquad (19)$$

In this representation, using the matrix elements defined by Eq. (16) and the expression for the dipolar coupling defined by Eq. (2a), specialized to the case of two interacting spins, the MAS dipolar Hamiltonian can be written in the form:

$$H_D(t) = \omega^D(t)[I_z^{12} - I_z^{34} - I_x^{23}] \qquad (20)$$

In this notation, the last term I_x^{23} represents the flip-flop term that is retained only in homonuclear spin systems.

2.3 Spin Evolution

The spin density operator $\rho(t)$ enables us to describe the evolution of nuclear spins during NMR experiments. The operator representing the state of the spin system satisfies the Liouville equation [54]:

$$\frac{d}{dt}\rho(t) = -i[H(t), \rho(t)], \tag{21}$$

which has the formal solution:

$$\rho(t) = U(t)\rho(0)U^{-1}(t), \tag{22}$$

where the evolution operator is given by:

$$U(t) = T\exp\left\{-i\int_0^t H(t')dt'\right\} \tag{23}$$

All relaxation phenomena are omitted from these equations. The Dyson time-ordering operator T necessitates a stepwise propagation to evaluate the time-evolution operator, which makes analytic, as well as numerical, calculations difficult to perform. Because the complexity involved in calculating $\rho(t)$ depends strongly on the form of the Hamiltonian, various approaches have been used to solve the Liouville equation for the time-evolution of the density matrix. In this review, we discuss two of these approaches, namely, Floquet Theory and Average Hamiltonian Theory.

Our first example of spin evolution under magic angle spinning is the calculation of the free induction decay (FID) of a single spin having a non-vanishing CSA. The initial density matrix of this spin system can be written in its reduced form $\rho(0) = 2I_x$, which corresponds to unit transverse magnetization. The FID signal of the magnetization is given by the expression:

$$\begin{aligned} S(t) &= 2\mathrm{Tr}[U(t)I_x U^{-1}(t)I_-] \\ &= \exp\{-i\Omega^{\mathrm{CS}}(t)\}, \end{aligned} \tag{24}$$

where:

$$\begin{aligned} U(t) &= \exp\{-i\Omega^{\mathrm{CS}}(t)I_z\} \\ &= 2\exp\{-i\tfrac{1}{2}\Omega^{\mathrm{CS}}(t)I_z\}I_z^{11} + 2\exp\{+i\tfrac{1}{2}\Omega^{\mathrm{CS}}(t)\}I_z^{22} \end{aligned} \tag{25}$$

Because the CS Hamiltonian $H_{\mathrm{CS}}(t)$ is self-commuting at different times, the Dyson time-ordering operator T in Eq. (23) can be omitted from the calculation of the $U(t)$. Accordingly, the time-evolution operator is characterized solely by the phase angle of rotation $\Omega^{\mathrm{CS}}(t)$, which is obtained by integration of the coefficient of $H_{\mathrm{CS}}(t)$ in Eq. (10):

$$\Omega^{\mathrm{CS}}(t) = -\Delta\omega t - \omega^{\mathrm{CSA}}\int_0^t \{g_1\cos(\omega_{\mathrm{R}}t' + \psi_1) + g_2\cos(2\omega_{\mathrm{R}}t' + \psi_2)\}dt' \tag{26}$$

It is convenient to expand the matrix elements of $U(t)$ in a Fourier series, provided that the isotropic phase $\Delta\omega t$ is separated from the expansion. Integration of the Fourier expansion of the coefficient $\omega^{CS}(t)$ in Eq. (13) leads to the relationship:

$$\exp\{-i\tfrac{1}{2}\Omega^{CS}(t)\} = \exp\{i\tfrac{1}{2}\Delta\omega t\} \times$$
$$\exp\left\{-\sum_{n=-2,n\neq 0}^{2} \frac{-\omega_n}{n\omega_R}[\exp\{in\omega_R t\} - 1]\right\} \tag{27}$$

The Fourier series of the latter exponential can then be defined:

$$\exp\left\{-\sum_{n=2,n\neq 0}^{2} \frac{-\omega_n}{n\omega_R}[\exp\{in\omega_R t\} - 1]\right\} = \sum_{n=-\infty}^{\infty} \delta_n \exp\{in\omega_R t\} \tag{28}$$

An infinite set of complex sideband amplitude coefficients δ_n arises from the set of four Fourier modes ω_n appearing in the argument of the second exponential in Eq. (27). The most convenient way to obtain the leading amplitudes is by numerical computation. Their dependence on the coefficients ω_n is complicated, but in general, when $M\omega_R < |\omega^{CSA}g_{+1}|, |\omega^{CSA}g_{+2}| < N\omega_R$, the complex amplitudes δ_n are expected to be of significant magnitude in the range $M < n < N$.

Aside from the phase rotation $\Delta\omega t$ arising from the frequency offset, which is common to all crystallites in a powder sample, the time-evolution operator and the FID signal are periodic functions. For polycrystalline samples undergoing MAS, an important consequence of this periodicity is the refocusing of the FID signal and the occurrence of rotational echoes at multiples of the rotor period τ_R [20, 22, 55–57]. The Fourier transform of the FID, expressed in Eq. (24), results in a spectrum characterized by narrow spectral lines at positions $\Delta\omega + n\omega_R$ with intensities I_n, which are convolutions of the amplitudes defined by Eq. (28):

$$I_n(\alpha, \beta, \gamma) = \sum_{m=-\infty}^{\infty} \delta_m \delta_{n-m} \tag{29}$$

To facilitate the explicit evaluation of the spectral intensities $I_n(\alpha, \beta, \gamma)$, the expansion of the complex sideband amplitudes δ_n can be written in terms of Bessel functions of the parameters $(\omega_n/n\omega_R)$ [58, 59]. In the usual MAS experiment, the signal is obtained from a polycrystalline sample; therefore, the intensities of the centerband ($n = 0$) and sidebands ($n \neq 0$) of the MAS spectrum must be evaluated by integration over all possible crystallite orientations (indicated by the bar):

$$\bar{I}_n = \frac{1}{8\pi^2} \int_0^{2\pi} \int_{-1}^{1} \int_0^{2\pi} I_n(\alpha, \beta, \gamma)\, d\alpha\, d\cos\beta\, d\gamma \tag{30}$$

After orientational averaging over the rotational phase α, the intensities become real and positive, which implies absorptive spectral lines [60, 61]. Herzfeld and Berger [58] have provided a methodology for determining the CSA parameters

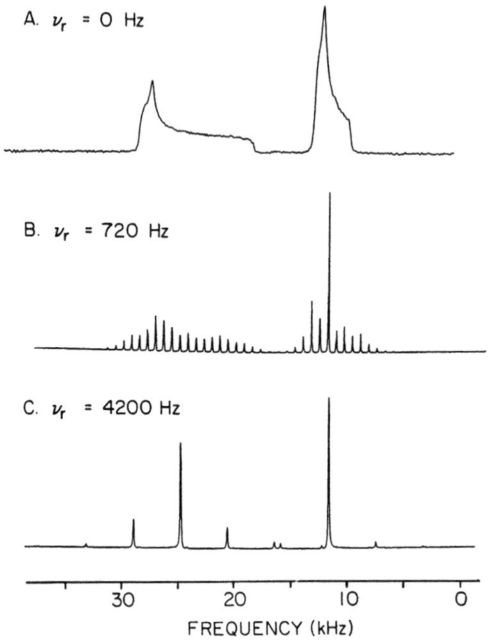

Fig. 2. ^{13}C NMR spectra of zinc acetate at 79.9 MHz. Reprinted with permission from Ref. [62]

from the observed centerband and sideband intensities. In Fig. 2, the ^{13}C spectrum of a polycrystalline sample of zinc acetate is shown with and without magic angle spinning [62]. At slow spinning speeds, the sideband intensities mirror the shape of the static powder spectrum [20, 55, 63], because the transient decay following each rotational echo tends to reflect the complete FID of the static powder before it reverses its time-evolution and forms the next echo. However, as the spinning speed is increased, the influence of the CSA is increasingly attenuated, leaving only the isotropic shifts at the centerbands.

2.4 *Effective Spin Hamiltonians*

The MAS Hamiltonian is periodic with a time constant τ_R. It is therefore not surprising that many experiments benefit by observation of the spin system synchronously with this time constant. Thus, in the discussion of MAS experiments which follows, we distinguish between situations where the data are acquired either asynchronously or synchronously with the period of mechanical rotation. In the first case, it is necessary to consider the full time-dependence of the Hamiltonian and determine $\rho(t)$ at all times. In the second, a time-independent average or effective Hamiltonian can be derived which provides a complete description of the stroboscopic behavior of $\rho(t)$.

Two theoretical approaches that make use of the periodicity of the Hamiltonian are Average Hamiltonian Theory (AHT) [29] and Floquet Theory [31–34].

AHT was initially developed to describe NMR experiments with pulse cycles modulo t_C (the cycle time), where the signal is acquired in synchrony with the pulse sequence, and it is therefore well-suited to describing cyclic MAS Hamiltonians [30, 49]. The AHT approach was utilized many years ago to obtain approximate effective Hamiltonians under MAS [22, 51]. Floquet Theory has also been applied to problems in magic angle spinning NMR, but less extensively [59, 64–67]. Only recently has it been demonstrated that this approach can be used to derive effective Hamiltonians for MAS experiments, which are also suitable for synchronous sampling [67]. However, in contrast to the AHT approach, the use of Floquet Theory allows a description of the evolution of the spin system at all times. It is therefore sometimes preferable to apply Floquet Theory as an alternative to the AHT method.

The utilization of the AHT and Floquet approaches leads to a better understanding of the time-evolution of nuclear spins primarily by providing effective Hamiltonians. Because of the periodicity of $H(t)$, the time-evolution operator satisfies the condition:

$$
\begin{aligned}
U(nt_C) &= T\exp\left\{-i\int_0^{nt_C} H(t)dt\right\} \\
&= \left[T\exp\left\{-i\int_0^{t_C} H(t)dt\right\}\right]^n \\
&= [U(t_C)]^n
\end{aligned}
\tag{31}
$$

If the spin system is monitored only at times $t = nt_C$, then the evolution operator $U(t_C)$ suffices to determine the spin evolution. An exact effective Hamiltonian $\bar{H}_{eff}$ can then be defined as the time-independent Hamiltonian satisfying the relationship [68–70]:

$$U(t_C) = \exp\{-i\bar{H}_{eff}t_C\} \tag{32}$$

2.4.1 The Toggling Frame

Several MAS recoupling experiments employ periodic sequences of RF pulses that, in effect, modify the time-dependent behavior of the spin Hamiltonian. Before either AHT or Floquet Theory is applied to a spin system governed by periodic interactions, it is often desirable to transform the system into a toggling frame [30] or jolting frame [71]. The toggling frame is particularly useful for the treatment of spin evolution during multiple pulse sequences in which the pulses are applied between points of signal acquisition.

The Hamiltonian $H^{tg}(t)$ in the toggling frame, just as the Hamiltonian in the usual rotating frame, must satisfy the constraint of periodicity, $H^{tg}(t + nt_C) = H^{tg}(t)$, where t_C is the length of the period. The transformation of the toggling frame is defined by the time-evolution operator for the RF fields [30]:

$$U_{\mathrm{RF}}(t) = T\exp\left\{-i\int_0^t H_{\mathrm{RF}}(t')dt'\right\}, \tag{33}$$

and the Hamiltonian in the toggling frame is defined by the expression:

$$H^{\mathrm{tg}}(t) = U_{\mathrm{RF}}^{-1}(t)[H(t) - H_{\mathrm{RF}}(t)]U_{\mathrm{RF}}(t) \tag{34}$$

In terms of the toggling frame Hamiltonian, the overall time-evolution operator can be expressed in the form [29, 72]:

$$U(t_C) = U_{\mathrm{RF}}(t_C)\left[T\exp\left\{-i\int_0^{t_C} H^{\mathrm{tg}}(t)dt\right\}\right] \tag{35}$$

The toggling frame transformation is most useful when the sequence of excitations is cyclic as well as periodic over a period t_C. When the cyclic condition $U_{\mathrm{RF}}(t_C) = 1$ is fulfilled, the evolution of the spins over the cycle is completely determined by the portion of the propagator involving $H^{\mathrm{tg}}(t)$:

$$\exp\left\{-i\bar{H}_{\mathrm{eff}}t_C\right\} = T\exp\left\{-i\int_0^{t_C} H^{\mathrm{tg}}(t)dt\right\} \tag{36}$$

If the RF pulse sequence is synchronous with the time-dependence of the sample rotation, then $t_C = N\tau_R$ for some integer N, and the MAS Hamiltonian remains periodic after transformation into the toggling frame. When the spin Hamiltonian is already time-dependent because of MAS, the toggling frame Hamiltonian $H^{\mathrm{tg}}(t)$ summarizes the combined modulations of the sample rotation and the RF pulses. Although the overall time-dependence of $H^{\mathrm{tg}}(t)$ is often quite complicated, Average Hamiltonian Theory and Floquet Theory provide valid approaches to the evaluation of the effective Hamiltonian defined by Eq. (36), provided that the time-dependent Hamiltonian remains periodic.

2.4.2 Average Hamiltonian Theory

To evaluate the effective Hamiltonian $\bar{H}_{\mathrm{eff}}$, the Magnus expansion [29, 73–75] can be applied:

$$\bar{H}_{\mathrm{eff}} = \bar{H}^{(0)} + \bar{H}^{(1)} + \bar{H}^{(2)} + \ldots, \tag{37}$$

where:

$$\bar{H}^{(0)} = \frac{1}{t_C}\int_0^{t_C} dt_1\, H(t_1);$$

$$\bar{H}^{(1)} = \frac{-i}{2t_C}\int_0^{t_C} dt_2\int_0^{t_2} dt_1 [H(t_2), H(t_1)];$$

$$\bar{H}^{(2)} = \frac{1}{6t_C}\int_0^{t_C} dt_3\int_0^{t_3} dt_2\int_0^{t_2} dt_1\, \{[H(t_3),[H(t_2),H(t_1)]] + [H(t_1),[H(t_2),H(t_3)]]\} \tag{38}$$

The zeroth and higher order terms in the expansion are all Hermitian. The zeroth order term is the average of $H(t)$ over the time period t_C. The calculation of this average is in many instances sufficient to obtain a reasonable estimate of the effective Hamiltonian $\bar{H}_{\text{eff}}$. In particular, if the Hamiltonian $H(t)$ commutes with itself at different times, then the first term $\bar{H}^{(0)}$ provides the exact result for $\bar{H}_{\text{eff}}$, since all terms of higher order involve commutators which vanish. Sometimes, symmetry considerations simplify the evaluation of higher order terms. For example, the odd order terms of the Magnus expansion vanish for Hamiltonians that are symmetric in time [30].

2.4.3 Floquet Theory

Floquet Theory [32, 33, 64] provides a completely general approach to solving the rate equation for the density matrix, given in Eq. (21), when the Hamiltonian is periodic in time. Here we use the description of Floquet approach as formulated by Shirley [31]. It is not our intention to derive the basic principles of the approach, but rather to present the methodology of its use and to introduce the tools necessary to apply it to MAS recoupling experiments. Derivations of the fundamental identities can be found elsewhere [65, 76].

The application of Floquet Theory involves the expansion of the periodic Hamiltonian in its Fourier series:

$$H(t) = \sum_{n=-\infty}^{\infty} H_n \exp\{in\omega_C t\}, \tag{39}$$

where we define the fundamental frequency $\omega_C = 2\pi/t_C$. The time-evolution operator in spin space is related to an equivalent time-evolution operator in Floquet space in the following way [31]:

$$\langle p|U(t)|q\rangle = \sum_{n=-\infty}^{\infty} \langle pn|U_F(t)|q0\rangle \exp\{in\omega_C t\}, \tag{40}$$

where the Floquet evolution operator $U_F(t)$ has the form:

$$U_F(t) = \exp\{-iH_F t\} \tag{41}$$

The matrix elements of the time-independent Floquet Hamiltonian operator H_F are defined as follows:

$$\langle pn|H_F|qm\rangle = n\omega_C \delta_{nm}\delta_{pq} + \langle p|H_{n-m}|q\rangle \tag{42}$$

The Floquet Hamiltonian H_F is a matrix of infinite dimensionality defined in the manifold of Floquet states $|pn\rangle$ with $n = -\infty, \ldots, \infty$. The Floquet states consist of spin states (labeled by the first index) dressed by the Fourier mode indices n of the time-dependent Hamiltonian.

The success of the Floquet approach depends upon the evaluation of the elements of $U_F(t)$. The difficulty in evaluating the Floquet evolution operator arises from the infinite dimensionality of the matrix representations. The advantage of the approach, however, is that the Dyson time-ordering operator is eliminated from the evolution operator in Floquet space, defined by Eq. (41). Depending on the form of H_F, the matrix elements of $U_F(t)$ can be calculated by exact diagonalization of the Floquet Hamiltonian H_F or approximated by the application of perturbation theory. In many cases, the Floquet approach yields exact or approximate solutions to the time-evolution of nuclear spins in NMR experiments, and concurrently, it often provides new physical insights.

The most important step in the solution of spin dynamics by Floquet Theory is the diagonalization of the time-independent Hamiltonian H_F. This diagonalization provides eigenvalues and eigenvectors satisfying the relationship:

$$H_F = D_F \Lambda_F D_F^{-1}, \tag{43}$$

which defines the diagonal matrix Λ_F and the matrix of eigenvectors D_F. The eigenvalues of the Floquet Hamiltonian can always be written in the form [31, 32]:

$$\begin{aligned} \lambda_n^p &= \langle pn | \Lambda_F | pn \rangle \\ &= \lambda^p + n\omega_C \end{aligned} \tag{44}$$

The eigenvectors $|\lambda_n^p\rangle$, which satisfy the relationship:

$$H_F |\lambda_n^p\rangle = \lambda_n^p |\lambda_n^p\rangle, \tag{45}$$

can be expressed in terms of the matrix elements of D_F:

$$d_{n-m}^{pq} = \langle pn | D_F | qm \rangle, \tag{46}$$

as a transformation of the original basis set of Floquet states:

$$|\lambda_n^p\rangle = \sum_q \sum_{m=-\infty}^{\infty} d_{m-n}^{qp} |qm\rangle \tag{47}$$

Knowledge of the diagonal form of H_F leads directly to the evaluation of nuclear spin dynamics. Using the Floquet eigenvalues and eigenvectors, an expression for the time-dependence of the relevant observable O can be derived [67]:

$$\begin{aligned} \mathrm{Tr}[\rho(t)O] = \sum_{p,q} \sum_{n=-\infty}^{\infty} &\left\{ \sum_{m=-\infty}^{\infty} \langle \lambda_0^q | R_F(0) | \lambda_m^p \rangle \right\} \\ &\langle \lambda_n^p | O_F | \lambda_0^q \rangle \exp\{i(\lambda^p - \lambda^q)t + in\omega_C t\} \end{aligned} \tag{48}$$

The intensity of each observed frequency $\{(\lambda^p - \lambda^q) + n\omega_C\}$ is proportional to the matrix element of O_F between the Floquet eigenstates $|\lambda_n^p\rangle$ and $|\lambda_0^q\rangle$, where the Floquet operator O_F corresponding to the observable of interest is defined by the matrix elements $\langle pn|O_F|qm\rangle = \langle p|O|q\rangle \delta_{nm}$. In addition, the intensity is scaled by a normalization factor for all $p-q$ transitions, which involves the initial density matrix $R_F(0)$ expressed in Floquet space. Its matrix elements are given by $\langle pn|R_F(0)|qm\rangle = \langle p|\rho(0)|q\rangle \delta_{nm}$. Through Eq. (48), the exact or approximate diagonalization of H_F provides the information necessary to calculate the complete time-dependence of the observable.

The spin Hamiltonian can be represented in terms of angular momentum operators having well-defined matrix elements in the manifold of spin eigenstates. To transform these matrices into their Floquet analogues, it is most convenient to utilize fictitious spin-half operators. To do so, Floquet operators are defined which correspond to the oscillating terms comprising the periodic Hamiltonian [34, 59]:

$$x_n \exp\{in\omega_C t\} I_x^{pq} \to x_n X_n^{pq};$$
$$y_n \exp\{in\omega_C t\} I_y^{pq} \to y_n Y_n^{pq};$$
$$z_n \exp\{in\omega_C t\} I_z^{pq} \to z_n Z_n^{pq} \quad (49)$$

The Floquet operators replace the fictitious spin-half operators in the spin Hamiltonian and eliminate the time-dependence of their coefficients. The Floquet operators have the following non-zero matrix elements:

$$\langle pn|X_{(n-m)}^{pq}|qm\rangle = \langle qn|X_{(n-m)}^{pq}|pm\rangle = \frac{1}{2};$$

$$\langle pn|Y_{(n-m)}^{pq}|qm\rangle = -\langle qn|Y_{(n-m)}^{pq}|pm\rangle = \frac{-i}{2}; \quad (50)$$

$$\langle pn|Z_{(n-m)}^{pq}|pm\rangle = -\langle qn|Z_{(n-m)}^{pq}|qm\rangle = \frac{1}{2}$$

The definitions of these operators deviate slightly from earlier definitions of the same operators [34, 59]. Schmidt and Vega have provided diagrams of these Floquet operator matrices, as well as several commutation relations obeyed by the Floquet operators [59]. In addition, the Floquet ladder operators have been introduced [67]:

$$P_n^{pq} = X_n^{pq} + iY_n^{pq};$$
$$M_n^{pq} = X_n^{pq} - iY_n^{pq}; \quad (51)$$

as well as the single block-diagonal operators Z_n^{pp} and N^{pp} with non-zero matrix elements:

$$\langle pn|Z_{(n-m)}^{pp}|pm\rangle = \tfrac{1}{2};$$
$$\langle pn|N^{pp}|pn\rangle = n \quad (52)$$

In the case of two states p and q, a time-dependent spin Hamiltonian of the form:

$$H(t) = 2\sum_{n=-\infty}^{\infty} (x_n \exp\{in\omega_C t\} I_x^{pq} + y_n \exp\{in\omega_C t\} I_y^{pq}$$
$$+ z_n \exp\{in\omega_C t\} I_z^{pq}), \quad (53)$$

is transformed into a time-independent Floquet Hamiltonian of the form:

$$H_F = \omega_C N^{pp} + \omega_C N^{qq} + 2 \sum_{n=-\infty}^{\infty} \{x_n X_n^{pq} + y_n Y_n^{pq} + z_n Z_n^{pq}\} \tag{54}$$

Alternatively, in terms of ladder and single block-diagonal operators, the Floquet Hamiltonian can be expressed:

$$H_F = \omega_C [N^{pp} + N^{qq}] + 2 \sum_{n=-\infty}^{\infty} \{z_n [Z_n^{pp} - Z_n^{qq}] + \tfrac{1}{2}(x_n - iy_n) P_n^{pq} + \tfrac{1}{2}(x_n + iy_n) M_n^{pq}\} \tag{55}$$

It is often useful to perform a partial diagonalization of the Floquet Hamiltonian, a procedure similar to the transformation of the usual spin Hamiltonian into the jolting frame [71]. If the Floquet Hamiltonian contains off-diagonal elements within its diagonal blocks Z_n^{pp} (meaning diagonal in the spin state index p, but not in the Fourier mode index n), then diagonalization of the blocks leads to a transformed Floquet Hamiltonian containing only the diagonal operators Z_0^{pp} and N^{pp}. These operators replace the blocks along the diagonal of the Floquet Hamiltonian matrix. At the same time, the transformation modifies the contents of the off-diagonal blocks, which involve the operators X_n^{pq} and Y_n^{pq}. However, these operators are not diagonalized. Thus, starting with the initial Floquet Hamiltonian:

$$H_F = \sum_p \omega_C N^{pp} + 2 \sum_{p<q} \sum_n \{z_n^{pq} Z_n^{pq} + x_n^{pq} X_n^{pq} + y_n^{pq} Y_n^{pq}\}, \tag{56}$$

the transformed Floquet Hamiltonian $H'_F = D_F^{-1} H_F D_F$ has a structure consisting only of block-diagonal operators N^{pp} and Z_0^{pp}, and off-diagonal operators X_n^{pq} and Y_n^{pq}.

The transformation matrix D_F is itself block-diagonal, and each component of the sum $D_F = \sum_p D_F^{pp}$ satisfies the relationship:

$$(D_F^{pp})^{-1} \left\{ \omega_C N^{pp} + 2 \sum_{n=-\infty}^{\infty} z_n^{pp} Z_n^{pp} \right\} (D_F^{pp}) = \omega_C N^{pp} + 2 z_0^{pp} Z_0^{pp} \tag{57}$$

Pursuing the derivation of Schmidt and Vega [59], the form of the transformation matrices D_F^{pp} for each block p is determined by applying the following commutation relations:

$$[Z_m^{pp}, Z_n^{pp}] = 0;$$
$$[N^{pp}, Z_n^{pp}] = n Z_n^{pp} \tag{58}$$

Using these relations, and the series expansion:

$$\exp\{iA\} B \exp\{-iA\} = B + i[A, B] + \tfrac{1}{2}[[A, B], A] + \ldots, \tag{59}$$

we obtain the following useful identities involving the operators Z_n^{pp} and N^{pp}:

$$\exp\{i\alpha Z_n^{pp}\}(z_m Z_m^{pp}) \exp\{-i\alpha Z_n^{pp}\} = z_m Z_m^{pp};$$
$$\exp\{i\alpha Z_n^{pp}\}(\omega_C N^{pp}) \exp\{-i\alpha Z_n^{pp}\} = \omega_C N^{pp} - in\alpha\omega_C Z_n^{pp} \tag{60}$$

The first relationship follows from the mutual commutation of the operators Z_n^{pp} and Z_m^{pp}. The second equation is also a consequence of this commutation relation, which leads to the vanishing of all terms beyond second order in the expansion given by Eq. (59).

Representing the unitary operator D_F^{pp} in the form of an exponential operator:

$$D_F^{pp} = \exp\{i\Delta_F^p\}, \tag{61}$$

we can employ the identities of Eq. (60) to deduce the general solution to Eq. (57). The result is that the argument assumes the form:

$$\Delta_F^p = i \sum_{n=-\infty, n\neq 0}^{\infty} \left(\frac{2z_n^{pp}}{n\omega_C} Z_n^{pp} \right) \tag{62}$$

When this result is inserted into the expression for D_F^{pp} in Eq. (61), we obtain the expression:

$$D_F^{pp} = \exp\left\{ - \sum_{n=-\infty, n\neq 0}^{\infty} \frac{2z_n^{pp}}{n\omega_C} Z_n^{pp} \right\} \tag{63}$$

The expression for D_F^{pp} can also be written in the more convenient form, using the eigenvector coefficients d_n^p:

$$D_F^{pp} = \sum_{n=-\infty}^{\infty} 2d_n^p Z_n^{pp} \tag{64}$$

The coefficients d_n^p satisfy a relationship in the time domain that is analogous to Eq. (28), which defined sideband amplitudes δ_n for the case of a single spin-half particle:

$$\exp\left\{ - \sum_{n=-\infty, n\neq 0}^{\infty} \frac{z_n^{pp}}{n\omega_C} [\exp\{in\omega_C t\} - 1] \right\} = \sum_{n=-\infty}^{\infty} d_n^p \exp\{in\omega_C t\} \tag{65}$$

It is again possible to express the coefficients d_n^p in terms of Bessel functions of the ratios $(z_n^{pp}/n\omega_C)$ [59] or to calculate them numerically. The relationship given by Eq. (65) can be demonstrated as follows: beginning with the time-dependent Hamiltonian corresponding to each independent diagonal block p:

$$H(t) = \sum_{n=-\infty}^{\infty} 2z_n^{pp} \exp\{in\omega_C t\} I_z^{pp}, \tag{66}$$

the evolution operator is obtained straightforwardly:

$$U(t) = \exp\{-iz_0^{pp}t\} \exp\left\{ - \sum_{n=-\infty, n\neq 0}^{\infty} \frac{z_n^{pp}}{n\omega_C} [\exp\{in\omega_C t\} - 1] \right\} \cdot 2I_z^{pp};$$

$$\langle p|U(t)|p\rangle = \exp\{-iz_0^{pp}t\} \exp\left\{ - \sum_{n=-\infty, n\neq 0}^{\infty} \frac{z_n^{pp}}{n\omega_C} [\exp\{in\omega_C t\} - 1] \right\} \tag{67}$$

Returning to Floquet space, we evaluate the same matrix element using Eq. (40):

$$\begin{aligned}
\langle p|U(t)|p\rangle &= \sum_{n=-\infty}^{\infty} \langle pn|U_F(t)|p0\rangle \exp\{in\omega_C t\} \\
&= \sum_{n=-\infty}^{\infty} \sum_{m=-\infty}^{\infty} \langle pn|\lambda_m^p\rangle \exp\{-i(z_0^{pp} + m\omega_C)t\} \\
&\quad \langle \lambda_m^p|p0\rangle \exp\{in\omega_C t\} \\
&= \sum_{n=-\infty}^{\infty} \sum_{m=-\infty}^{\infty} d_{n-m}^p \exp\{-i(z_0^{pp} + m\omega_C)t\} d_{-m}^{p*} \exp\{in\omega_C t\} \\
&= \exp\{-iz_0^{pp}t\} \sum_{n=-\infty}^{\infty} \sum_{m=-\infty}^{\infty} d_{n-m}^p d_{-m}^{p*} \exp\{i(n-m)\omega_C t\} \\
&= \exp\{-iz_0^{pp}t\} \sum_{n=-\infty}^{\infty} d_n^p \sum_{m=-\infty}^{\infty} d_{-m}^{p*} \exp\{in\omega_C t\}
\end{aligned} \tag{68}$$

The Floquet eigenvalues are given by $(z_0^{pp} + m\omega_R)$. Since the eigenvector coefficients are normalized, $\sum_{n=-\infty}^{\infty} d_{-n}^{p*} = 1$, and comparison of Eqs. (67) and (68) leads to Eq. (65), which provides an explicit relationship obeyed by the components d_n^p.

To obtain an effective Hamiltonian, the Hamiltonian in Eq. (56) is block-diagonalized:

$$\begin{aligned}
H'_F &= \sum_p \{\omega_C N^{pp} + 2z_0^{pp} Z_0^{pp}\} \\
&\quad + 2\sum_{p<q} \sum_{n=-\infty}^{\infty} (D_F^{pp})^{-1} \{x_n^{pp} X_n^{pq} + y_n^{pq} Y_n^{pq}\} (D_F^{qq})
\end{aligned} \tag{69}$$

This Z-diagonal Hamiltonian can be implemented in place of the original Hamiltonian for all synchronously detected observations. This result follows from the fact that the elements $d_n^p = \langle pm + n|D_F|pm\rangle$ are normalized, which, leads to the following relationship, starting from the fundamental expressions for the evolution operator given by Eqs. (40) and (41):

$$\begin{aligned}
\langle p|U(Nt_C)|q\rangle &= \sum_{n=-\infty}^{\infty} \sum_{m=-\infty}^{\infty} \sum_{k=-\infty}^{\infty} \langle pn|D_F|pm\rangle \\
&\quad \langle pm|\exp\{-iH'_F \cdot Nt_C\}|qk\rangle \langle qk|D_F^{-1}|q0\rangle \exp\{in\omega_C \cdot Nt_C\} \\
&= \sum_{n=-\infty}^{\infty} \langle pn|\exp\{-iH'_F \cdot Nt_C\}|q0\rangle
\end{aligned} \tag{70}$$

At times $t = Nt_C$, all phase factors $\exp\{in\omega_C \cdot Nt_C\} = 1$ and thus can be eliminated. The elements $\langle pn|D_F|pm\rangle$ of the transformation sum to unity and likewise disappear from the expression. Consequently, at times $t = Nt_C$, the evolution operator can be evaluated using H'_F instead of H_F.

Diagonalization of the Floquet Hamiltonian provides a means of calculating the effective Hamiltonian defined by Eq. (32). In the context of AHT, the effective

Hamiltonian is approximated by the first term in the Magnus expansion. In Floquet theory, one must evaluate the matrix elements of $\exp\{-iH'_F t_C\}$ and sum over them according to Eq. (70) to obtain the stroboscopic evolution operator in spin space, from which an exact effective Hamiltonian $\bar{H}_{eff}$ may be derived [67–70]:

$$\begin{aligned}\langle p|U(t_C)|q\rangle &= \langle p|\exp\{-i\bar{H}_{eff}t_C\}|q\rangle \\ &= \sum_{n=-\infty}^{\infty} \langle pn|\exp\{-iH'_F t_C\}|q0\rangle \end{aligned} \qquad (71)$$

In most cases, however, an exact evaluation is impractical. Instead, perturbation theory [54] is used to obtain an approximate Hamiltonian through the approximate diagonalization of the Floquet Hamiltonian [32, 34, 59, 65, 66, 70]. If the off-diagonal elements of H'_F are small relative to the differences between their corresponding diagonal elements, then zeroth order perturbation theory justifies the elimination of all off-diagonal elements except for those connecting nearly degenerate states [64]. Often, the off-diagonal elements are neglected after a block-diagonalization with respect to the dominant interactions in the Hamiltonian. The effective Hamiltonian $\bar{H}_{eff}$ is then constructed only from the diagonal interactions and those off-diagonal perturbations that link levels of nearly equivalent frequency in Floquet space.

Examples of the use of perturbation theory for the diagonalization of Floquet Hamiltonians can be found in several applications to MAS experiments [51, 64–66]. Although the approximate diagonalization procedure is useful in constructing effective Hamiltonians, the estimated eigenvalues and eigenvectors can also be used to calculated the behavior of observables at all times using Eq. (48). The approximate application of Floquet Theory is therefore not restricted to the case of synchronous sampling.

3 The Heteronuclear Dipolar Interaction

The attenuation of the dipolar interaction in MAS spectra greatly increases the spectral resolution of coupled spin systems. Concurrently, suppression of $H_D(t)$ makes it difficult to deduce the important structural information manifest in the dipolar coupling. This shortcoming has provided the impetus for the development of pulse sequences designed to reintroduce the dipolar interaction into the MAS experiment, and to separate and correlate this interaction with the chemical shift tensors of the spin system. In this section we present an overview of several techniques that are concerned with the recoupling of the heteronuclear dipolar interaction into MAS experiments. In general, these experiments are designed for isolated pairs of spins, I and S, that are coupled by the dipole-dipole interaction and that in addition experience anisotropic chemical shifts.

3.1 Isolated Heteronuclear Spin Pair under MAS

In the case where two interacting nuclei I and S have different gyromagnetic ratios, the dipolar MAS Hamiltonian given in Eq. (2a) is truncated in the presence of a high magnetic field, yielding the form:

$$H_D(t) = 2\omega_{IS}^D(t) I_z S_z \tag{72}$$

Like $H_{CS}(t)$, $H_D(t)$ is periodic modulo τ_R, and it is self-commuting. As a result, only the spin eigenvalues, but not the eigenstates, are perturbed during the rotor cycle, and the heteronuclear coupling generates rotational echoes, which lead to narrow centerbands and sidebands in the MAS spectrum [22]. The breadth of the dipolar sideband spectrum is of the order of the interaction strength. For rotation frequencies exceeding the coupling, the spectrum of a heteronuclear spin system is governed solely by the sideband pattern of the chemical shift interaction. Because the heteronuclear dipolar Hamiltonian commutes with the chemical shift Hamiltonian, all spectral features arising from the dipolar interaction can be evaluated independently and convoluted with the influence of the CS interactions for comparison with experiments.

3.1.1 The Dipolar MAS Spectrum

The MAS Hamiltonian for two spins can be represented in the manifold of spin states:

$$|1\rangle = |\alpha_I \alpha_S\rangle; \quad |2\rangle = |\alpha_I \beta_S\rangle;$$
$$|3\rangle = |\beta_I \alpha_S\rangle; \quad |4\rangle = |\beta_I \beta_S\rangle; \tag{73}$$

and, in terms of fictitious spin-half operators, is expressed as follows:

$$\begin{aligned} H(t) = & -\omega_I^{CS}(t)[I_z^{13} + I_z^{24}] - \omega_S^{CS}(t)[I_z^{12} + I_z^{34}] \\ & + (\omega_{IS}^D(t) + \tfrac{1}{2}\omega_{IS}^J)[I_z^{12} - I_z^{34}] \end{aligned} \tag{74}$$

All terms in the Hamiltonian commute, and it is therefore possible in many cases to simplify the form of the Hamiltonian according to the characteristics of the experiment. For example, if we wish to evaluate the FID signal of the S spin, then the relevant Hamiltonian assumes the simpler form:

$$\begin{aligned} H(t) = & (-\omega_S^{CS}(t) + \omega_{IS}^D(t) + \tfrac{1}{2}\omega_{IS}^J)[I_z^{11} - I_z^{22}] \\ & + (-\omega_S^{CS}(t) - \omega_{IS}^D(t) - \tfrac{1}{2}\omega_{IS}^J)[I_z^{33} - I_z^{44}] \end{aligned} \tag{75}$$

The chemical shift of the I spin can be omitted from this expression because it commutes with the observed S spin magnetization as well as the other terms in the Hamiltonian. Therefore, it has no influence on the evolution of the signal.

Likewise, the evolution operator corresponding to the Hamiltonian of Eq. (75) can be factored into terms corresponding to commuting elements of the Hamiltonian. In addition, since the complete Hamiltonian commutes with itself at different times, the Dyson time ordering operator T can be omitted from the propagator:

$$U(t) = \exp\left\{ -i\int_0^t H(t')dt' \right\} = U_J(t)\,U_D(t)\,U_{CS}(t), \tag{76}$$

where:

$$\begin{aligned} U_{CS}(t) &= 2u_{CS}(t)[I_z^{11} + I_z^{33}] + 2u_{CS}^*(t)[I_z^{22} + I_z^{44}]; \\ U_D(t) &= 2u_D(t)[I_z^{11} + I_z^{44}] + 2u_D^*(t)[I_z^{22} + I_z^{33}]; \\ U_J(t) &= 2u_J(t)[I_z^{11} + I_z^{44}] + 2u_J^*(t)[I_z^{22} + I_z^{33}]; \end{aligned} \tag{77}$$

using the definitions:

$$\begin{aligned} u_{CS}(t) &= \exp\{i\tfrac{1}{2}\Delta\omega^S t\}\exp\left\{ -\sum_{m=-2,m\neq 0}^{2} \frac{-\omega_m^S}{m\omega_R}[\exp\{im\omega_R t\} - 1] \right\} \\ &= \exp\{i\tfrac{1}{2}\Delta\omega^S t\} \sum_{n=-\infty}^{\infty} \delta_n^S \exp\{in\omega_R t\}; \\ u_D(t) &= \exp\left\{ -\sum_{m=-2,m\neq 0}^{2} \frac{\omega_m^{IS}}{m\omega_R}[\exp\{im\omega_R t\} - 1] \right\} \\ &= \sum_{n=-\infty}^{\infty} \varepsilon_n^{IS} \exp\{in\omega_R t\}; \\ u_J(t) &= \exp\{-i\tfrac{1}{4}\omega_{IS}^J t\} \end{aligned} \tag{78}$$

We also define dipolar centerband and sideband intensities:

$$I_n^D(\theta,\phi) = \sum_{m=-\infty}^{\infty} \varepsilon_m^{IS}\varepsilon_{n-m}^{IS}, \tag{79}$$

which are fully analogous to the CS intensities found in the MAS spectra of single spins:

$$I_n^{CS}(\alpha,\beta,\gamma) = \sum_{m=-\infty}^{\infty} \delta_m^S \delta_{n-m}^S \tag{80}$$

The intensities of the individual crystallites in powder samples are functions of the initial Euler angles (α, β, γ) of the principal axis system of the S spin CSA tensor and the polar angles (θ, ϕ) of the internuclear vector with respect to the rotor frame. These sets of angles are not independent, and their relationship is determined by the orientation of the internuclear vector in the PAS of the CSA interaction [5, 24, 77]. The observed signal from a powder sample involves an average over all possible orientations of a common molecular reference system. For each orientation of the chosen frame with respect to the rotor, a set of Euler angles relating the PAS of each spin interaction to the rotor coordinate system can be determined. It is often most convenient to select the PAS of one of the spin interactions as the common reference frame. For instance, in the present case, the PAS of the S spin CSA can serve as a reference; with this choice, the calculation of the spectrum involves an average over all possible orientations (α, β, γ) of the CSA frame, and each set of Euler angle (α, β, γ) fixes the appropriate polar angles (θ, ϕ) of the internuclear vector for that crystallite.

The free induction decay of the S spin after cross polarization, beginning with an initial density matrix $\rho(0) = S_x = [I_x^{12} + I_x^{34}]$, can be obtained by calculating the coefficient of the component of $S_- = [I_x^{12} - iI_y^{12} + I_x^{34} - iI_y^{34}]$ in the density matrix as a function of time, as in Eq. (24). After the substitution of the expressions of Eq. (78) into $U(t)\rho(0)U^{-1}(t)$, we obtain the powder-averaged FID, neglecting relaxation:

$$\mathrm{Tr}\{\rho(t)S_-\} = \exp\{i\Delta\omega^S t\} \sum_{n=-\infty}^{\infty} \sum_{m=-\infty}^{\infty} \big(\tfrac{1}{2}\overline{I_m^{CS} I_{n-m}^{D}} \exp\{-i\tfrac{1}{2}\omega^J t\} + \tfrac{1}{2}\overline{I_m^{CS} I_{m-n}^{D*}} \exp\{+i\tfrac{1}{2}\omega^J t\}\big) \exp\{in\omega_R t\} \tag{81}$$

Its Fourier transform exhibits positive absorption mode spectral lines [60, 61] at $(\Delta\omega^S - \frac{1}{2}\omega^J + n\omega_R)$ and $(\Delta\omega^S + \frac{1}{2}\omega^J + n\omega_R)$ with intensities $\frac{1}{2}\sum_m \overline{I_m^{CS} I_{n-m}^{D}}$ and $\frac{1}{2}\sum_m \overline{I_m^{CS} I_{m-n}^{D*}}$, respectively [78]. In cases where there is no anisotropy in the S spin chemical shift tensor, only the centerband of the CS spectrum remains, implying that $I_0^{CS} = 1$, and the remaining MAS spectrum becomes symmetric about the isotropic chemical shift $\Delta\omega^S$ with amplitudes $\overline{I_n^D}$ and $\overline{I_{-n}^{D*}}$ at frequencies $(\Delta\omega^S - \frac{1}{2}\omega^J + n\omega_R)$ and $(\Delta\omega^S + \frac{1}{2}\omega^J + n\omega_R)$, respectively. Figure 3 presents numerical simulations of the pure dipolar spectrum of a typical $^{13}C - {}^{1}H$ coupling at two spinning speeds. For comparison, the simulation of the static spectrum

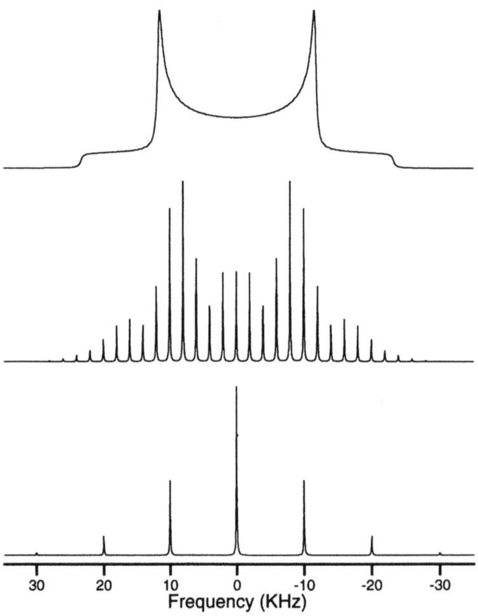

Fig. 3. Numerical simulations of heteronuclear dipolar spectra associated with a spin pair at 0 kHz, 2 kHz, and 10 kHz spinning speeds, respectively, from top to bottom. The chosen coupling constant is 23.3 kHz, which corresponds to a ^{1}H-^{13}C spin pair with 1.09 A separation, and all chemical shift interactions are omitted for simplicity. The static spectrum exhibits the symmetric Pake pattern corresponding to an isotropic distribution of internuclear vectors between the two spins. Under MAS conditions, the envelope of sideband intensities reflects the static powder pattern at slow spinning speeds. At greater spinning speeds, however, the influence of the interaction is increasingly attenuated, and the centerband predominates

is also shown. It exhibits the appropriate Pake pattern [1] for the heteronuclear dipolar coupling, which results from the inhomogeneous distribution of internuclear vector orientations in a powder sample. As in the analogous case of chemical shifts, illustrated by the spectra in Fig. 2, the sideband intensities in the MAS spectrum reflect the shape of the static powder pattern in the limit where the spinning is slow relative to the magnitude of the dipolar interaction.

In a heteronuclear spin pair, dipolar evolution during the free induction decay can be represented as a fictitious "magnetization" vector which evolves in a plane defined by $\langle S_x \rangle$ and $\langle 2S_yI_z \rangle$, the expectation values of the two coherences spanning the density matrix. After each τ_R, the vector returns to its initial position [16]. The periodic trajectories are analogous to those followed by the magnetization vector in the case of an isolated spin experiencing its CSA [51, 79, 80]. The application of π pulses to one of the spins alters the pathways of dipolar evolution and disrupts the formation of rotational echoes, as illustrated by the single crystallite trajectories shown in Fig. 4. In REDOR experiments [15, 16, 39], dephasing is promoted by the disruption of the dipolar trajectory by π pulses.

Because a signal that is influenced by the dipolar interaction generates rotational echoes after integer multiples of the rotor period, refocusing of the

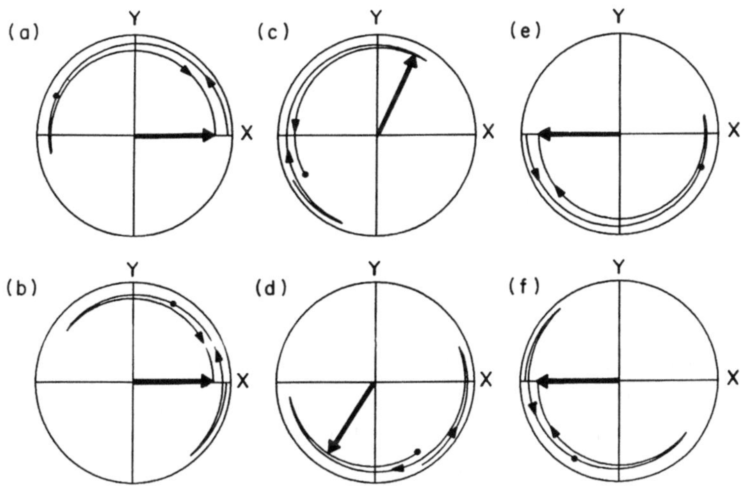

Fig. 4a–f. Calculated single crystallite trajectories under MAS and π pulses. The dipolar coupling constant is 5 kHz, the spinning frequency is 2 kHz, and $\theta = 60°$. **a–b** Dipolar evolution over one rotor cycle in a plane defined by $X \Leftrightarrow \langle S_x \rangle$ and $Y \Leftrightarrow \langle 2S_yI_z \rangle$ is shown for crystallites with $\phi = 60°$ and $\phi = 20°$, respectively. The cyclic nature of the trajectory reflects the formation of rotational echoes. The *dot* indicates the position of the "magnetization" vector at $\tau_R/2$, while the *long arrows* represent the orientation of the vector after multiples of τ_R. **c–d** The same crystallites are followed from $\tau_R/2$ to $3\tau_R/2$ after the application of a π pulse with phase X at $\tau_R/2$ applied to the *S* spin. The refocusing is disrupted. However, a second π pulse with phase Y at $3\tau_R/2$ reverses the time-evolution, leading to an echo at $2\tau_R$. **e–f** The evolution is followed from $2\tau_R$ until $3\tau_R$. These simulations were originally performed to examine the CSA case, but they are equally applicable to the dipolar coupling. Reprinted from Ref. [81] with permission

entire signal can be accomplished by a single π pulse at a rotational echo position, again as in the case of a single spin-half nucleus experiencing its CSA [51]. Such pulses are employed in 2D experiments in which the chemical shift is eliminated in one of the frequency domains. These experiments separate the dipolar sideband pattern from the shift anisotropy pattern [24–27, 81, 82]. In the regime where $\omega_R \ll |H_{CS}|$ and $\omega_R \ll |H_D|$, there are significant sideband intensities associated with both interactions, and 2D experiments can provide the relative orientations of the two interaction tensors through correlation of their sideband intensities [24, 83].

3.1.2 The Dipolar Floquet Hamiltonian

Neglecting chemical shifts, the combined heteronuclear dipolar and J-coupling Hamiltonian of a spin pair has the form:

$$H^{IS}(t) = (\omega_D^{IS}(t) + \tfrac{1}{2}\omega_J^{IS})[I_z^{11} - I_z^{22} - I_z^{33} + I_z^{44}], \tag{82}$$

further simplifying the Hamiltonian of Eq. (75). With the use of the Floquet operators defined by Eq. (52), and the transformation of the Hamiltonian into Floquet space defined by Eq. (49), the heteronuclear interaction can be represented in its Floquet form:

$$H_F^{IS} = \omega_R[N^{11} + N^{22} + N^{33} + N^{44}] + \sum_{n=-\infty}^{\infty} 2\omega_n^{IS}[Z_n^{11} - Z_n^{22} - Z_n^{33} + Z_n^{44}] \tag{83}$$

The coefficients ω_n^{IS} of the overall spin-spin interaction are the Fourier components of the dipolar interaction coefficients for $n = \pm 1, \pm 2$ and the J-coupling frequency $\omega_0^{IS} = 1/4\omega_{IS}^J$ for $n = 0$. All other Fourier components vanish. However, for simplicity, we generally omit the J-coupling from our further discussion of the heteronuclear spin pair and focus on the role of the much larger dipolar coupling.

Using the technique of block-diagonalization, we transform the Floquet Hamiltonian into its diagonal representation:

$$D_F^{-1} H_F^{IS} D_F = \omega_R[N^{11} + N^{22} + N^{33} + N^{44}] + 2\omega_0^{IS}[Z_0^{11} - Z_0^{22} - Z_0^{33} + Z_0^{44}] \tag{84}$$

In the case of pure dipolar coupling, the $n = 0$ Fourier coefficient ω_0^{IS} is vanishing, and therefore only the operators N^{pp} remain in the diagonalized Hamiltonian. The Floquet eigenvalues are simply $n\omega_R$. The matrix D_F consists of a sum of four block-diagonal transformations:

$$D_F = \sum_{n=-\infty}^{\infty} 2e_n^{IS} Z_n^{11} + \sum_{n=-\infty}^{\infty} 2e_{-n}^{IS*} Z_n^{22} + \sum_{n=-\infty}^{\infty} 2e_{-n}^{IS*} Z_n^{33} + \sum_{n=-\infty}^{\infty} 2e_n^{IS} Z_n^{44}, \tag{85}$$

where the coefficients e_n^{IS} are obtained in a way similar to the derivation of the elements d_n^p, which were defined in Eq. (64). Each term of D_F diagonalizes the corresponding block-diagonal operator in Eq. (83). The coefficients e_n^{IS} obey a relationship similar to that of Eq. (65):

$$\exp\left\{-\sum_{m=-\infty,m\neq 0}^{\infty}\frac{\omega_m^{IS}}{m\omega_R}[\exp\{im\omega_R t\}-1]\right\}=\sum_{n=-\infty}^{\infty}e_n^{IS}\exp\{in\omega_R t\} \quad (86)$$

Comparison of this identity with Eq. (78) reveals that $e_n^{IS}=\varepsilon_n^{IS}$.

The diagonal Hamiltonian of Eq. (84) can be utilized when the data are collected stroboscopically with the period of sample rotation. The Hamiltonian in Eq. (83) consists solely of Floquet operators Z_n^{pp} that are diagonal in the spin state p, and it contains no $n=0$ Fourier components. Therefore, the effective Hamiltonian for the case of stroboscopic sampling is exact and simple: it predicts no spin evolution at multiples of the rotor period, corresponding to the formation of complete rotational echoes. In other words, the effective Hamiltonian for synchronous observation is vanishing following each rotor cycle, even though transient behavior does occur between rotational echo positions. The behavior on a short time scale is reflected in the sideband intensities observed in the MAS spectrum. In more general circumstances, the same transformation D_F is applicable to the partial diagonalization of Hamiltonians containing operators which are not block-diagonal, such as RF fields and the flip-flop terms that arise in coupled homonuclear spin systems.

In the case of the heteronuclear dipolar Hamiltonian of Eq. (82), the Floquet eigenvectors are given by the following states:

$$|\lambda_n^1\rangle=\sum_{=-\infty}^{\infty}e_{m-n}^{IS}|1m\rangle;\quad |\lambda_n^2\rangle=\sum_{m=-\infty}^{\infty}e_{n-m}^{IS*}|2m\rangle;$$

$$|\lambda_n^3\rangle=\sum_{m=-\infty}^{\infty}e_{n-m}^{IS*}|3m\rangle;\quad |\lambda_n^4\rangle=\sum_{m=-\infty}^{\infty}e_{m-n}^{IS}|4m\rangle \quad (87)$$

Since all of the eigenvalues are of the form $m\omega_R$, the observed transitions also occur at frequencies $n\omega_R$, and the total intensities of the centerband and sidebands of the purely dipolar MAS spectrum can be evaluated straightforwardly, beginning with Eq. (48):

$$\frac{1}{2}(I_n^D+I_{-n}^{D*})=\frac{1}{2}\sum_{m=-\infty}^{\infty}(e_m^{IS}e_{n-m}^{IS}+e_{-m}^{IS*}e_{m-n}^{IS*}) \quad (88)$$

The result is consistent with Eq. (81), the expression derived directly in the time domain. The addition of the CS interaction to the Floquet Hamiltonian modifies the coefficients of the Z_n^{pp} operators; however, the eigenvalues and eigenvectors can still be derived straightforwardly through the use of the diagonalization parameters d_n^p. The effect of the J-coupling is to split each line in the MAS spectrum into two of equal intensity, separated by the J-coupling constant ω^J.

3.2 Rotary Resonance Recoupling

The continuous irradiation of one type of spin, while monitoring another set of spins, is a common technique in NMR spectroscopy [10]. For strong RF fields, the irradiation causes decoupling [72, 84], but for weak fields recoupling can occur in MAS experiments [35]. Experiments involving the weak irradiation of spin pairs in rotating solids have been discussed extensively by Levitt et al. [36]. They showed that the narrow centerband and sidebands of the usual MAS spectrum exhibit broadening when the RF field strength is a multiple of the rotor frequency, $\Omega^{\mathrm{I}}_{\mathrm{RF}} = N\omega_{\mathrm{R}}$. An appropriate pulse sequence for this experiment is illustrated in Fig. 5.

The line-broadening results from the recoupling of the dipolar interaction into the MAS spectrum. This effect is of particular significance when the dipolar interaction is smaller than the spinning frequency, for in such cases the influence of the dipolar coupling on the sideband intensities in the MAS spectrum is too small to utilize as a means of determining the dipolar coupling strength. The physical basis of the dipolar recoupling effect is the interaction between the modulation of the dipolar coupling by MAS, on the one hand, and the modulation of the non-observed spin I_z operator by the RF fields, on the other, in a suitable interaction frame [36]. When the rotary resonance condition $\Omega^{\mathrm{I}}_{\mathrm{RF}} = N\omega_{\mathrm{R}}$ is satisfied, destructive interference occurs between the two contributions to the time-dependence, and the result is an effective coupling whose amplitude is the resonant Fourier mode $|\omega^{\mathrm{IS}}_{-\mathrm{N}}|$ of the dipolar coupling.

More rigorously, the behavior of the heteronuclear spin pair at rotary resonance can be explained by Floquet Theory [37]. When the I spin is irradiated and the FID of the S spin observed, the relevant Floquet Hamiltonian assumes the form:

$$\begin{aligned} H_{\mathrm{F}} = {} & \omega_{\mathrm{R}}[N^{11} + N^{22} + N^{33} + N^{44}] \\ & - \sum_{n=-2}^{2} 2\omega^{\mathrm{S}}_{n}[Z^{11}_{n} - Z^{22}_{n} + Z^{33}_{n} - Z^{44}_{n}] \\ & + \sum_{n=-2}^{2} 2\omega^{\mathrm{IS}}_{n}[Z^{11}_{n} - Z^{22}_{n} - Z^{33}_{n} + Z^{44}_{n}] + \Omega^{\mathrm{I}}_{\mathrm{RF}}[X^{13}_{0} + X^{24}_{0}], \end{aligned} \tag{89}$$

where the continuous RF field on the I spin has the form $\Omega^{\mathrm{I}}_{\mathrm{RF}}[X^{13}_{0} + X^{24}_{0}]$,

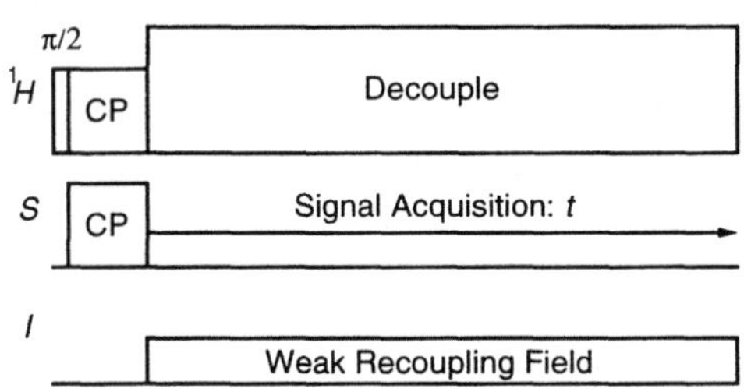

Fig. 5. Pulse sequence used to observe the MAS spectrum of an S spin coupled to an I spin, which is irradiated with a weak RF field. Cross polarization and proton decoupling are also employed in order to observe the FID of the dilute S spin in the triple resonance experiment

which contributes an off-diagonal perturbation to the Floquet Hamiltonian. Because the RF field is time-independent in the rotating frame, its representation as a Floquet operator consists only of an $n = 0$ Fourier component. The CS Hamiltonian of the I spin is omitted for simplicity. In order to evaluate the FID of the S spin during the irradiation of the I spin, the Floquet Hamiltonian must be diagonalized. This procedure can be carried out numerically, and it results in exact eigenvalues that differ from the original eigenvalues corresponding to the Z-diagonal part of the Hamiltonian. These shifts arise from the presence of the off-diagonal RF irradiation term.

For a weak dipolar interaction, the perturbed eigenvalues can be estimated by an approximate diagonalization involving three steps. The first transformation is the diagonalization of the S spin CS Hamiltonian. Because the CS terms of the S spin commute with the rest of the Hamiltonian, the first diagonalization affects neither the RF term nor the dipolar interaction in the Hamiltonian, and we obtain the result:

$$\begin{aligned} D_{F1}^{-1} H_F D_{F1} = {} & \omega_R [N^{11} + N^{22} + N^{33} + N^{44}] \\ & - \Delta\omega^S [Z_0^{11} - Z_0^{22} + Z_0^{33} - Z_0^{44}] \\ & + \sum_{n=-2}^{2} 2\omega_n^{IS} [Z_n^{11} - Z_n^{22} - Z_n^{33} + Z_n^{44}] + \Omega_{RF}^{I} [X_0^{13} + X_0^{24}] \end{aligned} \tag{90}$$

The RF field amplitude Ω_{RF}^{I} can be larger than the spinning frequency ω_R, so we next diagonalize the dominant terms X_0^{pq} and apply perturbation theory through zeroth order to estimate the eigenvalues and eigenvectors. Accordingly, the second transformation converts the Floquet Hamiltonian into the form:

$$\begin{aligned} D_{F2}^{-1} D_{F1}^{-1} H_F D_{F1} D_{F2} = {} & \omega_R [N^{11} + N^{22} + N^{33} + N^{44}] \\ & - \Delta\omega^S [Z_0^{11} - Z_0^{22} + Z_0^{33} - Z_0^{44}] \\ & + \Omega_{RF}^{I} [Z_0^{11} + Z_0^{22} - Z_0^{33} - Z_0^{44}] \\ & + \sum_{n=-2}^{2} \{\omega_n^{IS} [P_n^{13} - P_n^{24}] + \omega_{-n}^{IS*} [M_n^{13} - M_n^{24}]\}, \end{aligned} \tag{91}$$

where we utilize the fact that $\omega_n^{IS} = \omega_{-n}^{IS*}$.

For small dipolar coefficients $\omega_n^{IS} \ll \omega_R$, the off-diagonal terms of this Hamiltonian can be neglected under most circumstances. However, when some of these elements connect nearly degenerate elements along the diagonal, they become significant. This situation arises at the rotary resonance condition $\Omega_{RF}^{I} = N\omega_R$ for $N = \pm 1, \pm 2$. Figure 6 illustrates the general form of the Floquet Hamiltonian matrix in Eq. (91) with $\Delta\omega^S = 0$. Figure 7 shows a small portion of the blocks involving $p, q = 1, 3$ in greater detail. Only the elements connecting nearly degenerate states along the diagonal of Eq. (91) must be retained, and

$$\begin{bmatrix} H_F^{11} & 0 & H_F^{13} & 0 \\ 0 & H_F^{22} & 0 & H_F^{24} \\ H_F^{31} & 0 & H_F^{33} & 0 \\ 0 & H_F^{42} & 0 & H_F^{44} \end{bmatrix} \Rightarrow \begin{bmatrix} H_F^{11} & H_F^{13} \\ H_F^{31} & H_F^{33} \end{bmatrix}, \begin{bmatrix} H_F^{22} & H_F^{24} \\ H_F^{42} & H_F^{44} \end{bmatrix}$$

Fig. 6. Illustration of the blocks in the Floquet Hamiltonian matrix corresponding to Eq. 91. The matrix consists of two independent submatrices that are each 2 × 2 in the spin state index *p*. Each block H_F^{pp} is of infinite dimensionality in the Fourier mode index *n*

$$\left[\begin{array}{ccc|ccc} \boxed{+\omega_R+\frac{1}{2}\Omega^I_{RF}} & 0 & 0 & 0 & \omega_1^{IS} & \omega_2^{IS} \\ 0 & \boxed{+\frac{1}{2}\Omega^I_{RF}} & 0 & \boxed{\omega_{-1}^{IS}} & 0 & \omega_1^{IS} \\ 0 & 0 & \boxed{-\omega_R+\frac{1}{2}\Omega^I_{RF}} & \omega_{-2}^{IS} & \boxed{\omega_{-1}^{IS}} & 0 \\ \hline 0 & \boxed{\omega_1^{IS*}} & \omega_2^{IS*} & \boxed{+\omega_R-\frac{1}{2}\Omega^I_{RF}} & 0 & 0 \\ \omega_1^{IS*} & 0 & \boxed{\omega_1^{IS*}} & 0 & \boxed{-\frac{1}{2}\Omega^I_{RF}} & 0 \\ \omega_2^{IS*} & \omega_1^{IS*} & 0 & 0 & 0 & \boxed{-\omega_R-\frac{1}{2}\Omega^I_{RF}} \end{array}\right]$$

Fig. 7. Part of the Floquet Hamiltonian matrix in more detail. A small portion of the submatrix in Fig. 6 involving the blocks H_F^{pp} with $p = 1, 3$ is illustrated with Fourier modes arising from the diagonal operators N^{pp} in the range $n = +1, 0, -1$. The dominant interaction, namely the RF field, is placed along the diagonal, and the Fourier components of the dipolar coupling act as small off-diagonal perturbations. Near $N = 1$ rotary resonance, certain off-diagonal elements, which are *boxed*, connect nearly degenerate states. Only these terms and the diagonal of the matrix are retained in the subsequent approximate diagonalization

the approximate Hamiltonian takes the form:

$$\begin{aligned} D_{F2}^{-1}D_{F1}^{-1}H_F D_{F1}D_{F2} \approx \omega_R[N^{11}+N^{22}+N^{33}+N^{44}] \\ -\Delta\omega^S[Z_0^{11}-Z_0^{22}+Z_0^{33}-Z_0^{44}] \\ +\Omega^I_{RF}[Z_0^{11}+Z_0^{22}-Z_0^{33}-Z_0^{44}] \\ +\omega^{IS}_{-N}[P^{13}_{-N}-P^{24}_{-N}]+\omega^{IS*}_{-N}[M^{13}_N-M^{24}_N] \end{aligned} \quad (92)$$

In the region of rotary resonance, degenerate perturbation theory [54] requires the further transformation of interacting states which lie at approximately equal levels.

The Hamiltonian is straightforward to diagonalize because there is only one significant off-diagonal element ω^{IS}_{-N} connecting pairs of nearly degenerate Floquet states. These levels become equivalent when $-\frac{1}{2}\Delta\omega^S+\frac{1}{2}\Omega^I_{RF}=N\omega_R-\frac{1}{2}\Delta\omega^S-\frac{1}{2}\Omega^I_{RF}$ and $+\frac{1}{2}\Delta\omega^S+\frac{1}{2}\Omega^I_{RF}=N\omega_R-\frac{1}{2}\Delta\omega^S-\frac{1}{2}\Omega^I_{RF}$. The final 2 × 2 diagonalizations are demonstrated in Fig. 8. At exact rotary resonance, $\Omega^I_{RF}=N\omega_R$,

$$\begin{bmatrix} n\omega_R+\frac{1}{2}\Omega^I_{RF} & \omega^{IS}_{-1} \\ \omega^{IS*}_{-1} & (n+1)\omega_R-\frac{1}{2}\Omega^I_{RF} \end{bmatrix}$$

$$\Downarrow$$

$$\begin{bmatrix} \left(n+\frac{1}{2}\right)\omega_R+\frac{1}{2}\sqrt{(\Omega^I_{RF}-\omega_R)^2+4|\omega^{IS}_{-1}|^2} & 0 \\ 0 & \left(n+\frac{1}{2}\right)\omega_R-\frac{1}{2}\sqrt{(\Omega^I_{RF}-\omega_R)^2+4|\omega^{IS}_{-1}|^2} \end{bmatrix}$$

$$\Downarrow$$

$$\begin{bmatrix} \left(n+\frac{1}{2}\right)\omega_R+|\omega^{IS}_{-1}| & 0 \\ 0 & \left(n+\frac{1}{2}\right)\omega_R-|\omega^{IS}_{-1}| \end{bmatrix}$$

Fig. 8. Final diagonalizations near $N = 1$ rotary resonance. The boxed elements in the matrix illustrated by Fig. 7 represent part of an infinite set of uncoupled 2×2 matrices, indexed by the contribution $n\omega_R$ to the diagonal from N^{PP}. The avoided degeneracies occur at $\Omega^I_{RF} = \omega_R$, where the term $|\omega^{IS}_{-1}|$ splits the levels

we obtain the approximate Floquet Hamiltonian:

$$\begin{aligned} H'_F \approx\; & \omega_R[N^{11} + N^{22} + N^{33} + N^{44}] - \Delta\omega^S[Z^{11}_0 - Z^{22}_0 + Z^{33}_0 - Z^{44}_0] \\ & + N\omega_R[Z^{11}_0 + Z^{22}_0 - Z^{33}_0 - Z^{44}_0] \\ & + 2|\omega^{IS}_{-N}|[Z^{11}_0 - Z^{22}_0 - Z^{33}_0 + Z^{44}_0], \end{aligned} \tag{93}$$

and the approximate eigenvalues of the Hamiltonian assume the form:

$$\begin{aligned} \lambda^1_n &= (n + \tfrac{1}{2}N)\omega_R - \tfrac{1}{2}\Delta\omega^S + |\omega^{IS}_{-N}|; \\ \lambda^2_n &= (n + \tfrac{1}{2}N)\omega_R + \tfrac{1}{2}\Delta\omega^S - |\omega^{IS}_{-N}|; \\ \lambda^3_n &= (n - \tfrac{1}{2}N)\omega_R - \tfrac{1}{2}\Delta\omega^S - |\omega^{IS}_{-N}|; \\ \lambda^4_n &= (n + \tfrac{1}{2}N)\omega_R + \tfrac{1}{2}\Delta\omega^S + |\omega^{IS}_{-N}| \end{aligned} \tag{94}$$

A diagram of the Floquet eigenvalues is illustrated in Fig. 9 in the region of rotary resonance.

It is immediately apparent that, instead of the usual MAS spectrum consisting of sharp lines separated by multiples of the spinning frequency, the spectrum of the irradiated system will exhibit a set of spectral lines broadened by the strength of the Fourier mode $4|\omega^{IS}_{-N}|$. Using the expression in Eq. (48), the transition amplitudes of the S spin FID can be calculated for all possible transition frequencies, which are the differences between the Floquet eigenvalues given by Eq. (94). The approximate result is that the signal consists of a centerband and sidebands that become doublets positioned at $(n\omega_R + \Delta\omega^S + 2|\omega^{IS}_N|)$ and $(n\omega_R + \Delta\omega^S - 2|\omega^{IS}_N|)$ with equal amplitudes. Thus, in polycrystal-

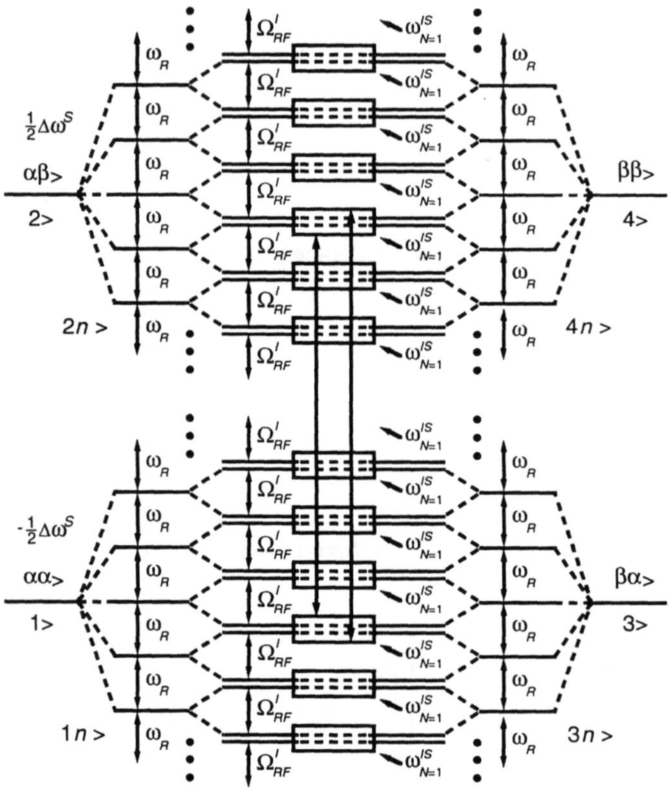

Fig. 9. Level diagram of the Floquet eigenvalues near $N = 1$ rotary resonance [37]. The time-dependence of MAS first splits the spin eigenvalues into a manifold spaced by ω_R. The RF field causes these states to experience avoided crossings near the rotary resonance conditions, where the weak heteronuclear coupling causes an inhomogeneous distribution of level shifts. The *vertical lines* connecting the final eigenvalues indicate prominent frequencies observed in the FID

line samples, instead of sharp sidebands, we obtain bands that broaden into narrow powder spectra because of the dependence of the coefficients $|\omega^{IS}_{-N}|$ on crystallite orientation.

The Hamiltonian of Eq. (93) can be implemented directly in the case of synchronous sampling. Expressed with the usual spin angular momentum operators, the effective Hamiltonian corresponding to H'_F has the form:

$$\bar{H}_{eff} = -\Delta\omega^S S_z + 2\bar{\omega}^D_{eff} I_z S_z, \tag{95}$$

at exact rotary resonance, where the effective dipolar frequency is given by:

$$\begin{aligned}\bar{\omega}^D_{eff} &= 2|\omega^{IS}_{-N}(\theta)| \\ &= \frac{\sqrt{2}}{4}\omega^D_{IS}\sin(2\theta)\end{aligned} \tag{96}$$

Powder spectra are generated by partial cancellation of the coherent averaging effect of MAS by the RF field. The line-broadening can in principle can be used to measure the dipolar interaction strength between the two spins. However, in practice, experimental difficulties associated with maintaining the narrow rotary resonance conditions and interpreting small lineshape perturbations limit the utility of the approach. The dipolar frequency depends on the orientational angle θ of the crystallite, but not on the phase angle ϕ, since it is only the magnitude of the Fourier coefficient defined by Eq. (8) which appears in the expression.

For simplicity, the treatment presented here neglects the CS interaction of the I spin, which does not commute with the rest of the Hamiltonian in the presence of the RF field. This non-commutation results in the appearance of weaker rotary resonances for $|N| > 2$. The effective couplings that are regenerated at the higher order resonance conditions are a strong function of the I spin chemical shift parameters. However, while the effects of higher order diminish with the inverse square of the spinning frequency, the more prominent splittings at $N = \pm 1, \pm 2$ are less dependent on the rate of rotation. More extensive discussions of the spectral broadening as a function of the resonance order N and of the deviation from the rotary resonance conditions have been presented by Levitt et al. [36] and Schmidt et al. [37].

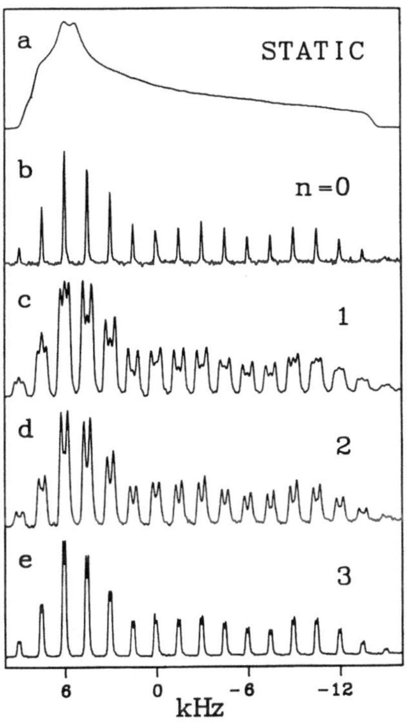

Fig. 10. ^{31}P Spectra of ^{15}N-labeled N-methyl-diphenylphosphoramidate. (*a*) The static powder pattern is dominated by the large CSA of the ^{31}P resonance. (*b*) In the absence of an RF field applied to ^{15}N, the spectrum breaks up into sharp sidebands with the application of MAS at the slow rotation rate of 1.5 kHz. (*c*)–(*e*) The spectral lines are broadened by the rotary resonance field, $\Omega^{I}_{RF} = N\omega_R$, which is applied on resonance to the ^{15}N spin under three conditions: $N = 1, 2, 3$. Note that the $N = 3$ resonance is much weaker than the others, since it arises from higher order effects involving the ^{15}N CSA. The regenerated powder patterns reflect the weak heteronuclear dipolar coupling of approximately 950 Hz. Reprinted from Ref. [35] with permission

In Fig. 10 ^{31}P MAS spectra broadened by weak irradiation are shown at several rotary resonance conditions [35]. In these spectra, the CSA is the predominant interaction. However, narrow powder patterns reflecting the weak $^{31}P - {}^{15}N$ dipolar coupling are reintroduced into the MAS spectrum. The $N = 1$ and $N = 2$ rotary resonance conditions lead to particularly strong recoupling, but a significant resonance effect is also observed at $N = 3$.

3.3 Dipolar Chemical Shift Correlation Spectroscopy

A variety of two dimensional (2D) MAS experiments have been introduced to examine the case of strong dipolar coupling between two spins, such as ^{13}C and ^{1}H [24–26, 38, 82, 83, 85–87]. In order to illustrate the essential ideas behind these techniques and to provide a sense of their diversity, the pulse sequences employed in just two of these experiments are shown in Fig. 11 [24, 80]. The design of the experiments takes advantage of the useful property that $H_D^{IS}(t)$ and $H_{CS}^{S}(t)$ commute at all times, even when π pulses are applied. The magnetizations of dilute S spins, such as ^{13}C or ^{15}N, are observed, and their interactions with I spins, generally ^{1}H, are examined through an analysis of the sideband intensities. The strong homonuclear interactions among I spins are usually suppressed during the dipolar part of the experiment using MREV-8 [88–90] or some other homonuclear decoupling scheme [91–93]. Homonuclear decoupling causes a scaling of the effective heteronuclear interaction, which must be accurately determined in order to interpret the experimental results. In all experiments, decoupling of the I spins from the S spin is achieved by strong heteronuclear decoupling during the signal acquisition, which constitutes the second time domain t_2 of the 2D experiment.

After excitation of the S spin coherence by cross polarization, the two time domains, t_1 and t_2, are characterized by different spin Hamiltonians $H_1(t_1)$ and $H_2(t_2)$. The evolution operators corresponding to the two Hamiltonians are $U_1(t_1)$ and $U_2(t_2)$, respectively, and the spin density matrix of the system exhibits

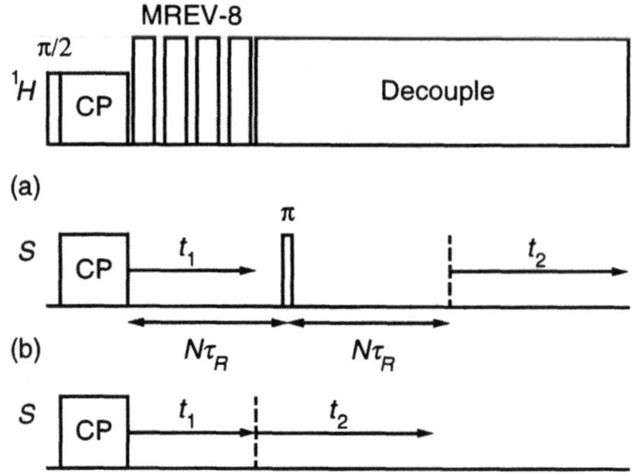

Fig. 11a, b. Two pulse sequences employed in 2D Dipolar Chemical Shift Correlation experiments. **a** In the first experiment, the chemical shift is refocused in the t_1 domain. **b** In the second sequence, the t_1 domain involves spin evolution under the influence of both the chemical shift and dipolar interaction the following time dependence during the 2D experiment:

$$\rho(t_1 + t_2) = U(t_1, t_2)\rho(0)U^{-1}(t_1, t_2);$$
$$U(t_1, t_2) = U_2(t_2)U_1(t_1) \tag{97}$$

In the first experiment illustrated in Fig. 11, the influence of the chemical shift term is eliminated in the t_1 domain by a rotor-synchronized π pulse applied in the middle of a constant time period. Because the dipolar coupling is active only during the first portion of the spin echo evolution period, it is not refocused by the π pulse. Therefore, the overall evolution operator is just $U_D(t_1)$ during the first part of the experiment, and a pure dipolar spectrum is obtained in the ω_1 dimension. During the t_2 period, heteronuclear decoupling is restored to the experiment, leaving only the chemical shift interaction to influence the spin evolution. Accordingly, the time-evolution operator can be expressed as follows:

$$U(t_1, t_2) = U_{CS}(t_2)U_D(t_1) \tag{98}$$

In the case of coupling to a single ^{1}H, the CS and dipolar sideband amplitudes, $I_n^{CS}(\alpha, \beta, \gamma)$ and $I_n^D(\theta, \phi)$, govern the intensities of the peaks in the two dimensional Fourier spectrum. After powder averaging, the frequencies and intensities of the MAS spectrum have the form:

$$(\omega_1, \omega_2) = (n\omega_R, \Delta\omega^S + m\omega_R); \quad I(\omega_1, \omega_2) = \tfrac{1}{2}\overline{(I_n^D + I_{-n}^{D*})I_m^{CS}} \tag{99}$$

The dipolar intensities must be calculated with a modified dipolar coupling constant determined by the homonuclear ^{1}H decoupling sequence.

Figure 12 demonstrates the separation of the dipolar from the CS interactions obtained in the 2D spectrum of ^{15}N-acetylvaline (NAV) [83]. The spectral lines can be fit to the two-dimensional sideband patterns in order to evaluate the dipolar interaction strength, as well as the orientation of the dipolar vector in the PAS frame of the S spin CSA tensor [24, 83]. This correlation is possible because of the dependence of the spectral intensities $I_n^{CS}(\alpha, \beta, \gamma)$ and $I_n^D(\theta, \phi)$ on both sets of Euler angles, which relate the principal axis systems of the spin

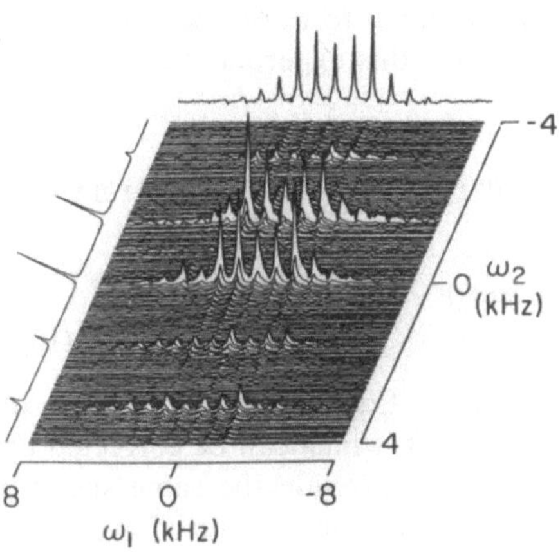

Fig. 12. 2D dipolar chemical shift correlation spectrum of ^{15}N-Acetylvaline. The FID of the labeled ^{15}N spin, which is coupled to a neighboring proton, is observed in the experiment. The pure dipolar spectrum obtained in ω_1 and the chemical shift spectrum obtained in ω_2 are reconstructed from projections of the 2D MAS spectrum. A spinning speed of 1.07 kHz was employed. Reprinted from Ref. [83] with permission

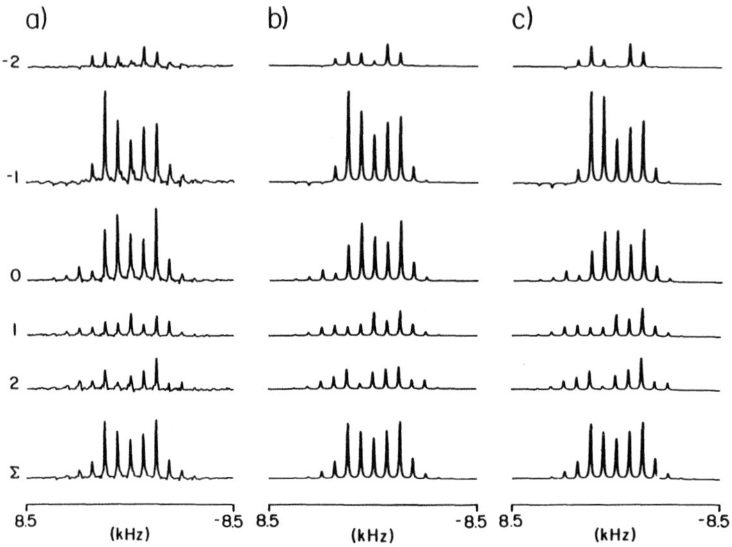

Fig. 13a–c. Dipolar cross sections from the 2D spectrum illustrated in Fig. 12 and computer simulations of the MAS spectrum. The slices are plotted parallel to the ω_1 domain through several chemical shift sidebands in ω_2. **a** Experimental spectrum. **b–c** Numerical simulations using two different orientations of the dipolar and chemical shift tensors. The two sets of polar angles, which define possible orientations of the internuclear vector in the CSA principal axis system of the observed spin, are (22°, 0°) and (17°, 0°), respectively. Reprinted from Ref. [83] with permission

interactions to the rotor system. In Fig. 13, slices of the same 2D spectrum, parallel to ω_1, are compared with simulations to illustrate the sensitivity of the spectral slices with respect to the relative orientations of the chemical shift and dipolar interaction tensors.

The second experiment in Fig. 11 is a simplified version of the first approach, which is modified so that the second time period immediately follows the first [80]. This experiment has the advantage that no delay period is used to refocus the chemical shifts, during which the signal would otherwise decay through relaxation. The result is an improvement of the signal-to-noise ratio of the 2D spectrum. The appropriate evolution operator for this experiment is given by:

$$U(t_1, t_2) = U_{CS}(t_1 + t_2)U_D(t_1) \tag{100}$$

The 2D MAS spectrum resulting from this time-evolution operator has spectral frequencies and intensities:

$$(\omega_1, \omega_2) = (\Delta\omega^S + (n+m)\omega_R, \Delta\omega^S + m\omega_R); \quad I(\omega_1, \omega_2) = \tfrac{1}{2}\overline{(I_n^D + I_{-n}^{D*})I_m^{CS}} \tag{101}$$

The difference between this result and that of Eq. (99) is just a shift of $(\Delta\omega^S + m\omega_R)$ in the ω_1 domain. Except for this shift, which can be corrected by a shearing transformation [80], both experiments provide the same spectral information. Other 2D experiments, by employing "mirror symmetric" evolution periods in the t_1 domain, exhibit peaks at frequencies $(\Delta\omega^S + n/2\,\omega_R, \Delta\omega^S + m\omega_R)$.

The dipolar peaks in these experiments appear at multiples of $1/2\omega_R$ with correspondingly greater sideband intensities, enhancing the influence of the dipolar coupling on the spectrum [25, 27, 94, 95].

An early demonstration of this class of experiments was the application of 2D dipolar chemical shift correlation techniques to the study of $^{15}N-{}^{1}H$ couplings in DNA by Diverdi and Opella [38]. They employed a pulse sequence similar to that of Figure 11(a), but homonuclear decoupling was accomplished by the use of the Lee-Goldberg technique [91], in which the protons are irradiated at the magic angle in spin space in order to suppress their homonuclear dipolar couplings. Their 2D spectrum, illustrated in Fig. 14 [38], shows the symmetrized spectra of the $^{15}N-{}^{1}H$ couplings resolved by the isotropic chemical shifts of the ^{15}N resonances. The experiment is able to distinguish among various possible protonation states in DNA, which are illustrated by Fig. 15.

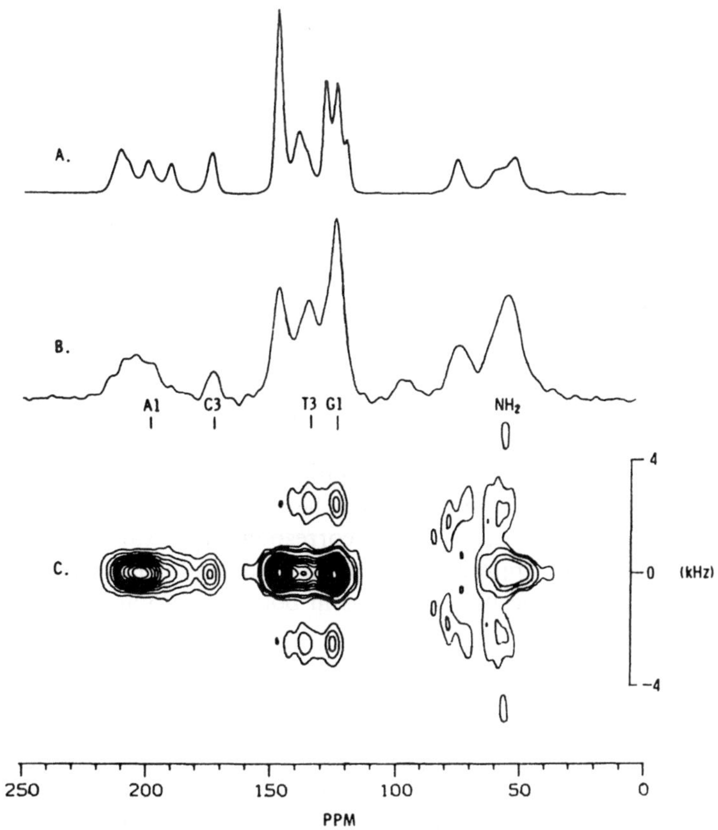

Fig. 14. 2D ^{15}N MAS separated local field spectrum of DNA. (*A*) Isotropic chemical shift spectrum of the hydrated B form of DNA (*B*) Isotropic chemical shift spectrum of low-humidity DNA (*C*) 2D spectrum showing dipolar coupling on the vertical axis and chemical shifts in the horizontal axis. The spectra were acquired at 15.24 MHz with a spinning speed of 2.2 kHz. Key ^{15}N resonances are labeled. Reprinted from Ref. [38] with permission

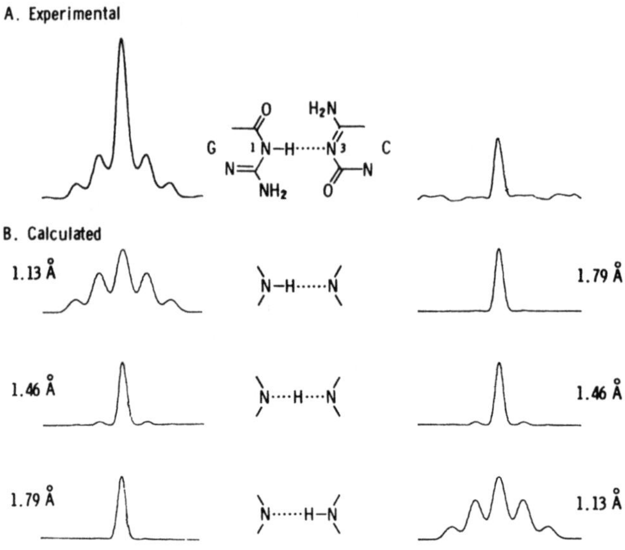

Fig. 15. Comparison of protonation states in DNA. (*A*) Experimental dipolar spectra obtained at chemical shifts corresponding to the nuclei in the illustration. (*B*) Comparison with simulations of possible protonation states. The experiments reveal that the G1 and T3 ^{15}N atoms have ^{15}N-^{1}H bond lengths of 1.1 Å. Reprinted from Ref. 38 with permission

Schaefer et al. have employed similar 2D experiments for the measurement of the dipolar interaction between ^{13}C and ^{1}H nuclei in isolated spin pairs [26]. Their approach, the Dipolar Rotational Spin Echo (DRS) experiment, utilizes the fact that for periodic self-commuting Hamiltonians the data measured at multiples of the period are identical when relaxation is neglected. The periodicity of the signal in the MAS experiment leads to a set of experiments in which t_1 is incremented by Δt for N steps such that $\tau_R = N\Delta t$. Fourier transformation in the t_1 domain then results in N frequency points for each line in the ω_2 spectrum. This procedure enables the discrimination among spin pairs with different motional characteristics, since the apparent dipolar coupling is an average over fast molecular motions. Various dipolar spectral patterns, which reflect the Pake powder lineshapes of the strong heteronuclear coupling, are obtained for effective dipolar interaction tensors corresponding to various types of motion. Applications of this approach have been reported for the study of fast orientational motions of molecular segments in polymers [86, 96–98] and biological molecules [99].

The 2D techniques discussed in this section were the first methods employed to isolate and measure heteronuclear dipolar couplings in rotating solids. However, these experiments differ in many respects from the others discussed in this review. In these 2D dipolar correlation experiments, the reintroduction of the dipolar interaction is achieved by the temporary removal of the heteronuclear decoupling field. Decoupling is necessary to eliminate very strong dipolar interactions even in the presence of magic angle spinning. With spin decoupling, the heteronuclear interaction, if it is substantially greater than the spinning speed, is clearly visible in the sideband structure of the dipolar MAS spectrum. This situation generally holds for the case of dilute spins coupled to protons, which possess a large gyromagnetic constant.

In weakly coupled spin pairs, however, the influence of the heteronuclear coupling is strongly attenuated at typical spinnig frequencies, and it becomes impossible to measure the dipolar coupling solely through its effect on the sideband intensities. For example, in the case a $^{13}C - {}^{15}N$ spin pair, the maximum coupling constant is less than 1 kHz, and it would be necessary to spin the sample much more slowly in order to observe significant dipolar sideband intensities. Instead, it is more practical to employ other experiments in which the effect of the dipolar coupling is deliberately magnified either by the enhancement of sideband intensities [25, 27, 94, 95] or by the outright interruption of dipolar echo formation [15, 16, 35, 36, 39]. Methods that operate by the dephasing of rotational echoes are not restricted to the slow spinning regime, where the signal-to-noise ratio is poor. The rotary resonance experiment achieves recoupling by this latter means. Under certain circumstances, namely, at the rotary resonance conditions, the application of the RF field to the non-observed spin prevents the refocusing of the dipolar interaction and leads to its effective reintroduction.

3.4 Rotational Echo Double Resonance

A more practical method for the study of weak dipolar couplings is the Rotational Echo Double Resonance (REDOR) experiment of Gullion and Schaefer [15, 16, 39], in which π pulses are employed to prevent the formation of dipolar rotational echoes. They introduced REDOR experiments in order to accurately determine the dipolar interaction strength between two dilute heteronuclear spins. These experiments combine the rotational refocusing of the CSA and heteronuclear dipolar interactions with the dephasing effects of π pulses on rotating samples [51, 81]. In static samples, the dipole-dipole interaction of a spin pair can be refocused by the application of a π pulse to one of the spins. However, simultaneously applied pulses to both spins do not affect the dipolar evolution. These effects were utilized in the development of Spin Echo Double Resonance (SEDOR) experiments for the detection of dipolar interactions in static samples [6, 10].

3.4.1 Dipolar Dephasing Experiments

For rotating samples, rotor-synchronized spin echo experiments have been introduced [15, 16], such as the one illustrated in Fig. 16 [100]. In this MAS experiment, the heteronuclear interaction is recoupled by the application of two π pulses per rotor period applied to either the I or S spin. REDOR experiments are designed to measure the dipolar interaction between heteronuclear spins other than protons. For example, the I and S spins can represent ^{13}C and ^{15}N,

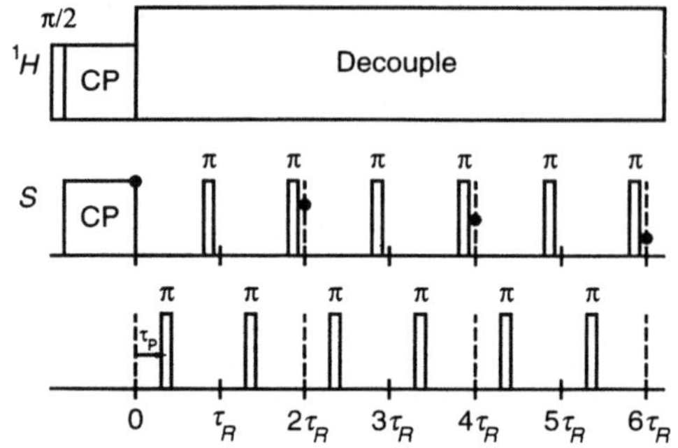

Fig. 16. REDOR pulse sequence. Two equally spaced π pulses are applied per rotor period with a variable time delay τ_P after cross polarization [16, 39, 100]. The *large dots* indicate measurement of the signal, either in a series of experiments or by acquisition with sampling windows

or other dilute spin pairs. In these experiments, the detected spin is first excited by proton cross polarization. Following excitation, the observed transverse magnetization evolves under the influence of the dipolar and J-coupling interactions, the CS interactions of the two spins, and the π pulses applied to the system. The location of the two π pulses within each rotor period can be varied, resulting in various expressions for the recoupled dipolar interaction [101]. In all REDOR experiments, the combined effect of the two π pulses is to hinder the refocusing of the dipolar interaction under MAS and cause dephasing of the signal. The rate of dephasing reflects the magnitude of the dipole-dipole coupling. In the case illustrated by Fig. 16, the pulses are equally spaced, but the overall phase ψ of the pulse sequence is left variable.

In order to eliminate the influence of the CS interaction of the S spin on the overall spin evolution, one π pulse is applied per rotor period in order to refocus its chemical shift interaction, including the CSA. In the absence of dipolar dephasing, the result is the formation of a full spin echo and the recovery of the transverse signal after every two rotor periods [51]. Following the REDOR dephasing sequence, the FID of the S spin is acquired and transformed into its Fourier spectrum with varying periods of dephasing $N\tau_R$. This procedure yields the total spectral intensity $\overline{S(N\tau_R)}$ of the monitored S spin magnetization as a function of the REDOR evolution time $N\tau_R$.

Because the acquisition begins at rotational echo positions and is therefore synchronous with the rotor, an effective Hamiltonian for the REDOR pulse sequence can be constructed that characterizes the evolution of the spin pair. The π pulses are taken to be ideal and instantaneous rotations. The most convenient way to evaluate the effective Hamiltonian is to transform the dipolar coupling into the toggling frame and calculate it using AHT. However, it is also possible to apply Floquet Theory to the analysis of dipolar recoupling by REDOR [37]. The dephasing experiment consists of n periods of two rotor cycles each, so that there are $N = 2n$ rotor periods altogether in the dephasing interval. The first pulse is applied at time τ_P following cross polarization, corresponding to the phase angle $\psi = \omega_R \tau_P$ with respect to the rotor period.

The dipolar Hamiltonian assumes the following form in the toggling frame:

$$H_{D}^{tg}(t) = 2\omega_{IS}^{D}(t)F(t)I_{z}S_{z}, \tag{102}$$

where $F(t)$ accounts for the action of the π pulses in the toggling frame, which is to flip the sign of the dipolar coupling at each point where a pulse is applied. The function $F(t)$ is plotted in Fig. 17. A π pulse on either channel switches the sign of the toggling function $F(t)$, but does not modify the form of the spin operators.

A similar result applies to the chemical shift interactions. However, in the case of chemical shifts, the π pulses flip only the sign of the CS interaction associated with the spin to which the pulse is applied. Since in all cases the π pulses preserve the form of the spin operators, the dipolar and CS interactions of the two spins commute with one another in the toggling frame in exact analogy to the usual situation in the absence of π pulses. Consequently, the dipolar REDOR Hamiltonian is completely independent of the CS parameters, and only consideration of the heteronuclear interaction is required for its derivation. The commutation of the CS terms with the spin-spin coupling terms, which is retained under π pulse sequences, is also useful in the design of dipolar chemical shift correlation experiments, but it does not apply to more general pulse sequences, such as the continuous RF field employed in rotary resonance experiments. In the latter case, the CSA interaction of the I spin plays a major role in the spin dynamics at relatively low spinning speeds, implying that its parameters must be estimated in order to obtain the dipolar coupling from the observed lineshapes. Similar considerations apply to all of the homonuclear methods discussed in this review.

To calculate the AHT result for REDOR, the function $F(t)$, which defines the toggling frame transformation:

$$U_{RF}^{-1}(t)I_{z}S_{z}U_{RF}(t) = F(t)I_{z}S_{z}, \tag{103}$$

is written as a Fourier series:

$$F(t) = \sum_{n=-\infty}^{\infty} F_{n} \exp\{in\omega_{R}t\} \tag{104}$$

When combined with the expansion of $\omega^{D}(t)$, which was defined by Eqs. (7) and (8), the Fourier expansions facilitate the evaluation of the average Hamiltonian.

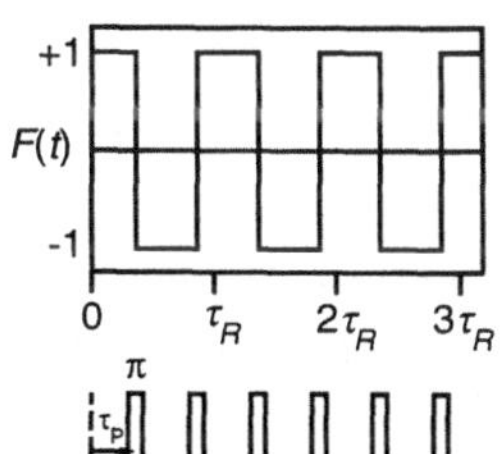

Fig. 17. Plot of the toggling function $F(t)$. The π pulses applied to either spin induce square wave modulation of the spin operator I_zS_z in the toggling frame

Because the Hamiltonian commutes with itself at different times in the toggling frame, the zeroth order term of the Magnus expansion provides the exact result for the effective Hamiltonian, regardless of the length of the cycle time. In addition, after a pair of π pulses has been applied to each spin, the toggling frame transformation is cyclic. Consequently, the AHT method can be straightforwardly applied, and the effective Hamiltonian assumes the form:

$$\begin{aligned} \bar{H}_{\mathrm{eff}} &= \sum_{n=-2}^{2} 4\omega_n^{\mathrm{IS}} F_{-n} I_z S_z \\ &= 2\bar{\omega}_{\mathrm{eff}}^{\mathrm{D}} I_z S_z \end{aligned} \tag{105}$$

The part of the average Hamiltonian corresponding to the CS interactions of the spins and their J-coupling vanishes over two rotor cycles. Therefore, the evolution of the transverse magnetization is determined only by the effective dipolar frequency $\bar{\omega}_{\mathrm{eff}}^{\mathrm{D}}$.

All Fourier coefficients ω_n^{IS} of $\omega_{\mathrm{IS}}^{\mathrm{D}}(t)$ are functions of the initial polar angles (θ, ϕ) of the dipolar vector in the rotor frame, and therefore the effective REDOR frequency $\bar{\omega}_{\mathrm{eff}}^{\mathrm{D}}$ depends upon the crystallite orientation. Likewise, the effective coupling depends upon the phase angle ψ associated with the RF pulse sequence [101], and is proportional to the dipolar interaction constant $\omega_{\mathrm{IS}}^{\mathrm{D}}$:

$$\bar{\omega}_{\mathrm{eff}}^{\mathrm{D}} = \frac{\sqrt{2}}{\pi} \omega_{\mathrm{IS}}^{\mathrm{D}} \sin 2\theta \sin(\phi + \psi) \tag{106}$$

In the particular case where $\psi = \pi$, the pulses are applied every $1/2\tau_{\mathrm{R}}$, and the dipolar frequency is given by:

$$\bar{\omega}_{\mathrm{eff}}^{\mathrm{D}} = \frac{-\sqrt{2}}{\pi} \omega_{\mathrm{IS}}^{\mathrm{D}} \sin 2\theta \sin \phi \tag{107}$$

For each crystallite in a powder sample, the dephasing trajectory follows the course:

$$\begin{aligned} S(N\tau_{\mathrm{R}}) &= \mathrm{Tr}[\exp\{-i\bar{H}_{\mathrm{eff}} N\tau_{\mathrm{R}}\} S_x \exp\{i\bar{H}_{\mathrm{eff}} N\tau_{\mathrm{R}}\} S_x] \\ &= \cos(\bar{\omega}_{\mathrm{eff}}^{\mathrm{D}} N\tau_{\mathrm{R}}) \end{aligned} \tag{108}$$

The ensemble-averaged signal $\overline{S(N\tau_{\mathrm{R}})}$ arising from all possible crystallite orientations in a powder sample depends explicitly on $\omega_{\mathrm{IS}}^{\mathrm{D}} N\tau_{\mathrm{R}}$ and can be expressed as a function of this parameter alone:

$$\overline{S(\omega_{\mathrm{IS}}^{\mathrm{D}} N\tau_{\mathrm{R}})} = \frac{1}{4\pi} \int_0^{2\pi} \int_{-1}^{+1} S(N\tau_{\mathrm{R}}) d\cos\theta \, d\phi \tag{109}$$

The inhomogeneous distribution of effective frequencies leads to the decay of the transverse magnetization. Determination of $\omega_{\mathrm{IS}}^{\mathrm{D}}$ from the observed dephasing of the signal enables the evaluation of the nuclear distance according to Eq. (4). Because of the mutual commutation among the terms of the spin Hamiltonian, prior knowledge or estimation of the CS parameters is not required to extract the dipolar coupling constant from the data.

The signal from a powder does not depend on the phase angle ψ of the pulse sequence because it serves only to add an additional constant to the phase angle ϕ associated with the dipolar vector of each crystallite. Although the dipolar frequency for any given crystallite is modified by the phase ψ, the signal from the isotropic distribution of crystallites remains unchanged after averaging over ϕ [101]. In addition, the CS interactions are refocused following every two rotor periods regardless of the angle ψ. Typically, the phase ψ is chosen to be π, so that the pulses are applied every half rotor cycle, alternating between the two spins [100]. However, in the case of TEDOR experiments, it is more convenient to apply the first pulse after one quarter of a rotor cycle, in which case $\psi = \pi/2$.

For the experimental determination of $\overline{S(\omega_{IS}^{D} N\tau_R)}$, the REDOR signal is monitored in conjunction with the reference signal $S_0(N\tau_R)$ obtained without the application of dipolar dephasing pulses to the I spin. Alternating detection of $\overline{S(\omega_{IS}^{D} N\tau_R)}$ and $S_0(N\tau_R)$ provides a relative difference function $\Delta S/S$ that is independent of the rate of transverse relaxation:

$$\frac{\Delta S}{S} = \frac{S_0(N\tau_R) - \overline{S(\omega_{IS}^{D} N\tau_R)}}{S_0(N\tau_R)} \tag{110}$$

The time-dependence of the REDOR difference function $\Delta S/S$, illustrated in Fig. 18 [39], is uniquely determined by the pulse sequence and the coupling constant ω_{IS}^{D}. Since uncoupled spins contribute to $\Delta S/S$ through the denominator

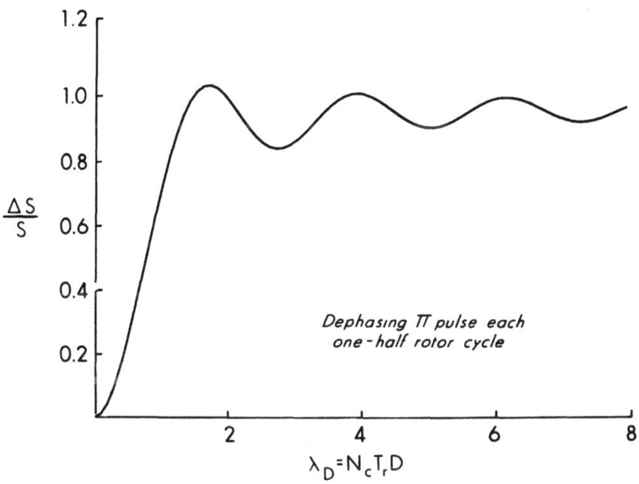

Fig. 18. Calculated REDOR difference curve as a function of the parameter $\omega_{IS}^{D} N\tau_R$. The time course of dephasing depends only on the dipolar coupling constant and the number of rotor cycles. Reprinted from Ref. [39] with permission

of Eq. (110), their relative abundance must be known in order to perform accurate distance measurements [102]. In addition, to obtain high accuracy, compensated spin echo sequences, most notably XY-8 [103], have been introduced and applied to REDOR experiments [104, 105]. These phase-alternated sequences greatly reduce the undesirable consequences of pulse imperfections, including RF inhomogeneity and resonance offset effects. The latter are particularly significant in the case of dilute spins, which often experience a broad range of chemical shifts. Precise synchronization of the rotor period with the pulse timings is also important, although the sequence in Fig. 16 with $\psi = \pi$ has been shown to minimize the deterioration of performance in spin echo formation when small timing errors are present [106].

Most REDOR measurements of internuclear distances have been performed on $^{13}C - {}^{15}N$ spin pairs [39, 107–109], but applications to other spin pairs have been reported [100, 110]. The REDOR method has also been extended to the case of coupling between a dilute spin-half nucleus and ^{2}H [111]. An experiment demonstrating the application of the REDOR technique to alanine is shown in Fig. 19 [39]. The REDOR difference ΔS removes uncoupled spins from the MAS spectrum and emphasizes the resonances of nuclei experiencing the largest interactions with the non-observed spins.

2D versions of the REDOR approach have been developed, which eliminate background signals from uncoupled spins [16]. Among them are Odd Dipolar

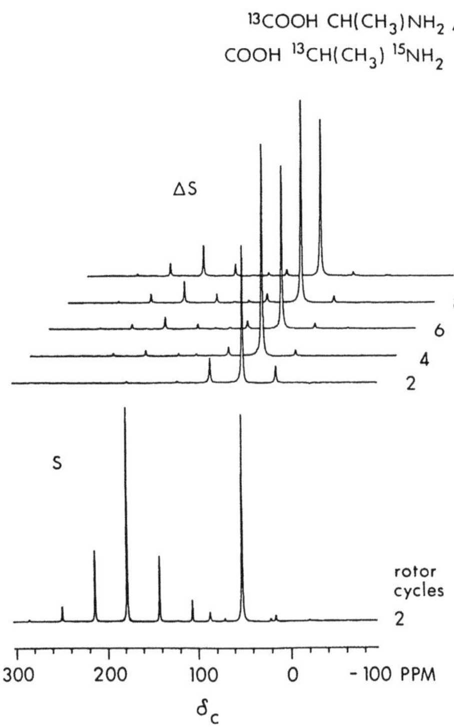

Fig. 19. ^{13}C REDOR spectra of an equimolar, recrystallized mixture of ^{13}C, ^{15}N-labeled L-alanines. The full signal S is shown at the bottom. In the ΔS spectrum, rapid dephasing is observed between the directly bonded nuclei, with a weaker signal developing at later times from the intermolecular coupling. The specta were acquired at 50.3 MHz with a spinning speed of 2.702 kHz. Reprinted from Ref. [39] with permission

Rotational Spin Echo (ODRSE) [95] and Extended Dipolar Modulation (XDM) experiments [94]. In these experiments, dipolar modulation is monitored by changing the pulse timings during an overall dephasing period whose length is fixed. Consequently, the relaxation of the transverse magnetization does not broaden the spectra in the time domain of dipolar evolution, and it is unnecessary to normalize the REDOR difference signal with $S_0(N\tau_R)$. In addition, Double REDOR and Rotational Echo Triple Resonance (RETRO) experiments have been proposed to eliminate background signals [112]. Double REDOR involves two REDOR difference experiments. The first step in the experiment is REDOR dephasing to a third heteronuclear spin, which distinguishes a selected resonance from signals arising from uncoupled spins. Following spin selection, a second REDOR dephasing experiment is performed to measure an internuclear distance without interference from unwanted signals contributing to $S_0(N\tau_R)$.

3.4.2 Transferred Echo Double Resonance

Another experimental approach that greatly attenuates the signals from uncoupled spins is based on the transfer of coherence from the I spin to the S spin using the heteronuclear coupling regenerated by REDOR. This coherence transfer is achieved by two $\pi/2$ pulses applied simultaneously to both spins at a rotational echo position, in analogy to the INEPT experiment of solution state NMR [113]. The basic Transferred Echo Double Resonance (TEDOR) experiment [114, 115] is illustrated in Fig. 20. After cross polarization of the I spin, dipolar evolution of the transverse magnetization under the influence of the REDOR sequence promotes the following evolution of I spin coherence:

$$\rho(N\tau_R) = I_x \cos(\bar{\omega}^D_{eff} N\tau_R) + 2I_yS_z \sin(\bar{\omega}^D_{eff} N\tau_R) \tag{111}$$

The magnetization dephases under the influence of the recoupled interaction through exchange with the non-observable single quantum coherence $2I_yS_z$. After N rotor cycles of spin evolution, the $2I_yS_z$ term is converted into $2I_zS_x$ by the pair of $\pi/2$ pulses, and the modified coherence evolves further under the

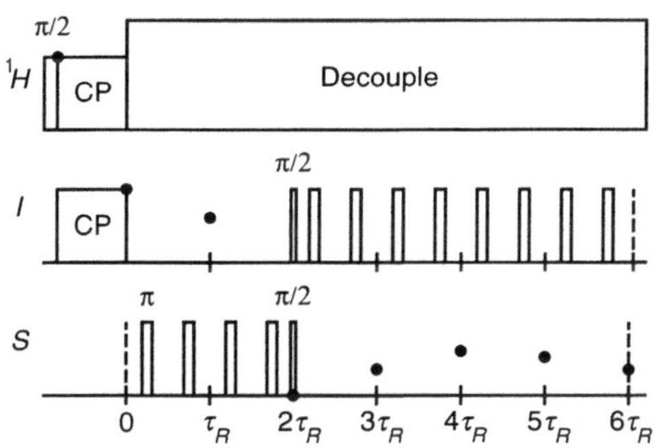

Fig. 20. Pulse sequence for the TEDOR experiment [114]. Magnetization begins on the I spin and evolves under two REDOR cycles until coherence transfer, after which further evolution is permitted for four cycles in this example. *Solid circles* indicate the pathway of observable magnetization

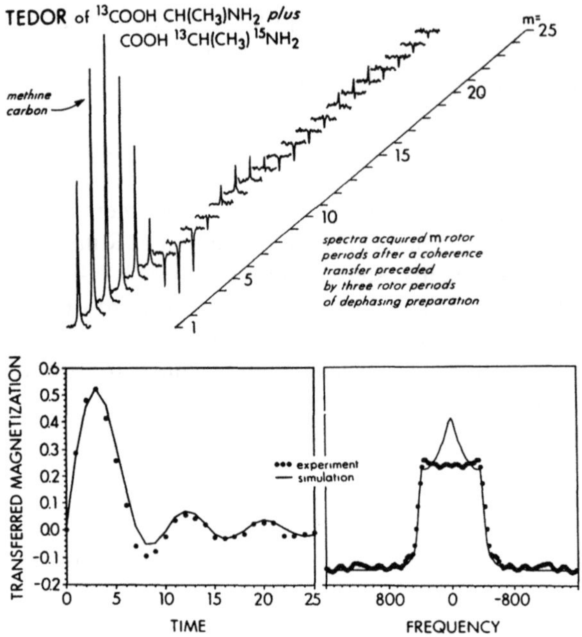

Fig. 21. Coherence transfer by TEDOR. With the same equimolar mixture of alanines shown in Fig. 19, the signal is exchanged from the ^{15}N the methine ^{13}C, but here the intermolecular carboxyl-^{13}C is efficiently filtered out because of its weaker coupling. Reprinted from Ref. [114] with permission

REDOR pulse cycle as follows:

$$\rho(N\tau_R, M\tau_R) = \sin(\bar{\omega}^D_{eff} N\tau_R)\{2I_z S_x \cos(\bar{\omega}^D_{eff} M\tau_R) + S_y \sin(\bar{\omega}^D_{eff} M\tau_R)\} \tag{112}$$

Overall, N rotor cycles are applied before the transfer and M rotor cycles following the $\pi/2$ pulses. The second term of Eq. (112) results in an observable S spin signal that is excited solely by the dipolar interaction and thus contains no contributions from the uncoupled S spins.

The detection of the S spin magnetization can begin after variable periods of evolution following the transfer of coherence in order to provide the maximal S spin signal intensity. The strength of the dipolar interaction determines the appropriate choice of parameters N and M. Figure 21 demonstrates experimental trajectories that are in agreement with numerical simulations. The TEDOR pulse sequence can also be used in the mixing time of 2D heteronuclear correlation experiments among dilute spins [116].

4 The Homonuclear Dipolar Interaction

The homonuclear dipolar interaction provides a direct route to internuclear distances among spins in solids possessing the same gyromagnetic constant γ. At the rotational resonance conditions, in particular, where the chemical shift difference $\Delta\omega = \Delta\omega_1 - \Delta\omega_2$ between two spins matches a multiple of the spinning frequency, $\Delta\omega = N\omega_R$, an effective dipolar interaction is restored to the MAS experiment, which can be used to interrogate internuclear distances [117–119]. In this section, we discuss the influence of the dipolar interaction among homonuclear spins on the NMR spectra of rotating samples and methods of recoupling these interactions into MAS experiments.

The distinction between the dipolar terms retained in the high field spin Hamiltonian of a homonuclear spin system and a heteronuclear spin system is the addition of the flip-flop interactions $[I_{+1}I_{-j} + I_{-i}I_{+j}]$. These terms do not commute with the operators I_{zi} of the individual spins i in the system and therefore cause the eigenfunctions of the Hamiltonian to become linear combinations of the product eigenfunctions of these operators. Under these conditions, selective irradiation of individual transitions becomes impossible, and one cannot, in general, "burn a hole" in the NMR spectrum of a system of coupled homonuclear spins [120]. In static solids, the homonuclear spin system is referred to as "homogeneous," in contrast to the heteronuclear spin system, which is "inhomogeneous" in the sense that a hole can be burned in its spectrum. In rotating solids, the amplitude modulation of the heteronuclear Hamiltonian makes the frequency-selective saturation of the spectrum impossible. However, because it commutes with itself at different times and leads to full rotational echoes, the heteronuclear Hamiltonian is still "inhomogeneous" in the MW sense [22]. In both the static and spinning cases, the inhomogeneous character of the Hamiltonian implies that signals can be refocused with suitable π pulse sequences [10, 51, 81].

In the general case, the various terms of the time-dependent MAS Hamiltonian containing flip-flop terms do not commute. Consequently, the homonuclear interaction is regarded as "homogeneous" in both static and rotating solids. With some special exceptions, the FID signal does not refocus after multiples of the rotor period, even in coupled pairs of spin-half nuclei [22]. Accordingly, the time-evolution operator fails to average to unity over a rotor cycle:

$$U(\tau_R) = T\exp\left\{-i\int_0^{\tau_R} H(t)dt\right\} \neq 1, \tag{113}$$

and the effective Hamiltonian is non-vanishing. Because the Hamiltonian is not self-commuting at different times, the Dyson time-ordering operator T must be considered explicitly in the evaluation of $U(\tau_R)$ and the corresponding effective Hamiltonian. In contrast, simple integration of the phase $\int_0^{\tau_R} H(t)dt$ is sufficient to evaluate the propagator in an entirely inhomogeneous spin system.

4.1 *Rotor Driven Recoupling*

Under most conditions, the chemical shift difference $\Delta\omega$ between the nuclei of a dilute homonuclear spin pair efficiently truncates the flip-flop term of a weak dipolar interaction, and the MAS spectrum resembles that of the heteronuclear

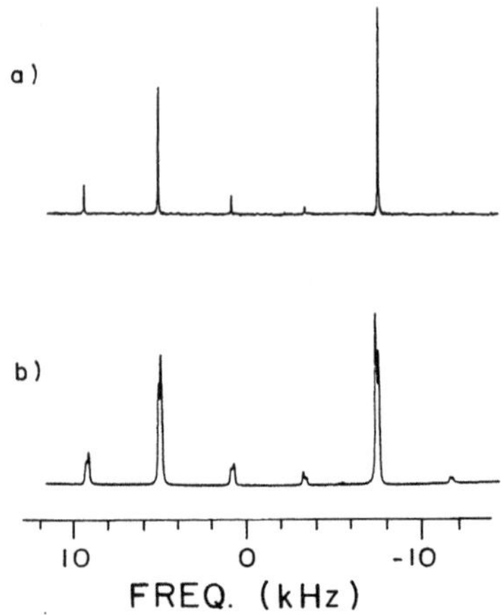

Fig. 22. Zinc acetate MAS Spectra. (*a*) Natural abundance spectrum at 4.38 kHz spinning speed, corresponding to $N = 3$ rotational resonance. (*b*) Spectrum of the doubly-^{13}C labeled sample, which demonstrates considerable line-broadening from the ≈ 2 kHz dipolar coupling. Reprinted from Ref. [40] with permission

case. However, when the centerband and sideband positions coalesce, line broadenings appear [117–119]. This recoupling effect, which is demonstrated by the MAS spectrum of zinc acetate in Fig. 22, occurs selectively at the rotational resonance (R^2) conditions [121].

Several theoretical approaches have been introduced to explain the spectral phenomena observed in MAS experiments on dilute homonuclear spin pairs. Maricq and Waugh [22] used Average Hamiltonian Theory to examine the case of ^{13}C-enriched diammonium oxalate, in which the chemical shift parameters of two ^{13}C spins are identical, but their CSA tensors differ in orientation. This case leads to a dipolar powder lineshape corresponding to rotational resonance with $N = 0$. To evaluate the lineshapes of the centerbands and sidebands in the more general case, Gan and Grant [122] and McDowell and co-workers [123–126] also employed Average Hamiltonian Theory, while Meier and Earl [119] followed an adiabatic approach to obtain the features of the spectrum in the limit of slow spinning. Numerical calculations and a theoretical model involving virtual states have been presented by Raleigh et al. [40, 118] and Levitt et al. [41]. The Floquet formalism has been applied by Schmidt and Vega [59], Nakai and McDowell [66, 127], and Kubo and McDowell [65] to the calculate the homonuclear MAS spectra involving two spins.

4.1.1 The Dipolar Floquet Hamiltonian

In this section, we discuss the application of Floquet Theory to the homonuclear spin pair [59, 66, 127]. We consider the general case in which there are significant CS differences between the spins. These interactions are taken to be the dominant terms in the Hamiltonian, and the dipolar flip-flop term contributes a small off-diagonal perturbation. An advantage of the Floquet approach is that it can be applied without the restriction of synchronous detection, which would otherwise limit the spectral window and cause aliasing in the evaluation of the NMR spectrum. However, for the more restrictive case of synchronous detection, Floquet Theory also provides an effective Hamiltonian that is useful in describing magnetization exchange experiments.

The Hamiltonian of a homonuclear spin pair rotating at the magic angle consists of the CS terms of the two spins, and their dipolar and J-couplings. As earlier, we omit the J-coupling for convenience, noting that it generally has a small influence in the experiments of interest in this discussion. However, there are some instances where the J-coupling is important in coupled homonuclear spins undergoing MAS [118], particularly the case of ^{31}P NMR spectroscopy [65, 123, 126, 128]. Asymmetric intensities in homonuclear MAS spectra provide evidence of the sign of J-coupling [127]. In addition, Challoner et al. [125] have proposed the possibility of observing the anisotropy of ^{31}P–^{31}P J-couplings in the MAS lineshapes of spin pairs.

To simplify the form of the Hamiltonian, we assume, without loss of generality, that the sum of the isotropic chemical shift values of the spins is vanishing. To treat homonuclear spin pairs, it is most convenient to define the Fourier components $\omega_n^{\pm}$ of the sums and differences of the chemical shift Hamiltonian. Accordingly, transforming the CS parameters of the two spins, we apply the following definitions:

$$\begin{aligned} &\Delta\omega = (\Delta\omega_1 - \Delta\omega_2); \quad (\Delta\omega_1 + \Delta\omega_2) = 0; \\ &\omega_n^{+} = (\omega_n^{1} + \omega_n^{2}); \quad \omega_n^{-} = (\omega_n^{1} - \omega_n^{2}) \end{aligned} \tag{114}$$

Again, we employ the spin states: $|1\rangle = |\alpha_1\alpha_2\rangle, |2\rangle = |\alpha_1\beta_2\rangle, |3\rangle = |\beta_1\alpha_2\rangle$, and $|4\rangle = |\beta_1\beta_2\rangle$. In this basis set, the Floquet Hamiltonian can be written in the form [59]:

$$\begin{aligned} H_F &= H_F^{11} + H_F^{23} + H_F^{44}; \\ H_F^{11} &= \omega_R N^{11} + \sum_{n=-2, n\neq 0}^{2} 2(-\omega_n^{+} + \omega_n^{D}) Z_n^{11}; \\ H_F^{44} &= \omega_R N^{44} + \sum_{n=-2, n\neq 0}^{2} 2(+\omega_n^{+} + \omega_n^{D}) Z_n^{44}; \\ H_F^{23} &= \omega_R (N^{22} + N^{33}) - \sum_{n=-2, n\neq 0}^{2} 2\omega_n^{D} [Z_n^{22} + Z_n^{33}] \\ &\quad - \Delta\omega [Z_0^{22} - Z_0^{33}] - \sum_{n=-2, n\neq 0}^{2} 2\omega_n^{-} [Z_n^{22} - Z_n^{33}] \\ &\quad - \sum_{n=-2, n\neq 0}^{2} \{\omega_n^{D} P_n^{23} + \omega_n^{D*} M_{-n}^{23}\} \end{aligned} \tag{115}$$

In this case, the block H_F^{23} is defined to include all operators A_n^{pq} with spin indices $p, q = 2, 3$. The last term of H_F^{23}, which involves the operators P_n^{23} and M_n^{23}, is the addition to the Hamiltonian of Eq. (83) representing the flip-flop term.

It is this term that complicates the diagonalization of H_F and causes the line-broadenings of the centerbands and sidebands near the rotational resonance conditions. Without the flip-flop contribution, the Floquet Hamiltonian is completely diagonal in the spin states p and Fourier modes n that constitute Floquet space. The result is sharp sideband structure in the MAS spectrum, reflecting the case of the heteronuclear spin system. The off-diagonal operators require the combined rotation of both the spin states and Fourier modes in order to achieve the diagonalization of H_F. This process leads to shifts in the lineshape positions that are different for each crystallite, and over a powder average, to inhomogeneous broadening in the NMR spectrum.

The independent blocks H_F^{11} and H_F^{44} of the Floquet Hamiltonian H_F can be diagonalized in a straightforward manner using the method of block-diagonalization, since they consist only of operators of the form Z_n^{pp} and N^{pp}. With the definition $\Delta\omega_1 + \Delta\omega_2 = 0$, the diagonalization of H_F^{11} and H_F^{44} leads to the results:

$$(D_F^{11})^{-1} H_F^{11} (D_F^{11}) = \omega_R N^{11};$$
$$(D_F^{44})^{-1} H_F^{44} (D_F^{44}) = \omega_R N^{44}; \tag{116}$$

where we once again employ transformations that render the submatrices H_F^{pp} diagonal:

$$D_F^{11} = \left\{ \sum_{n=-\infty}^{\infty} 2d_n^+ Z_n^{11} \right\} \left\{ \sum_{n=-\infty}^{\infty} 2e_n Z_n^{11} \right\} + 2Z_0^{22} + 2Z_0^{33} + 2Z_0^{44};$$
$$D_F^{44} = 2Z_0^{11} + 2Z_0^{22} + 2Z_0^{33} + \left\{ \sum_{n=-\infty}^{\infty} 2d_{-n}^{+*} Z_n^{44} \right\} \left\{ \sum_{n=-\infty}^{\infty} 2e_n Z_n^{44} \right\} \tag{117}$$

The coefficients d_n^+ and e_n are the CS and dipolar components of the Floquet transformation matrices, which are defined by the relationships:

$$\exp\left\{ - \sum_{n=-2, n\neq 0}^{2} \frac{-\omega_n^+}{n\omega_R} [\exp\{in\omega_R t\} - 1] \right\} = \sum_{n=-\infty}^{\infty} d_n^+ \exp\{in\omega_R t\};$$
$$\exp\left\{ - \sum_{n=-2, n\neq 0}^{2} \frac{\omega_n^D}{n\omega_R} [\exp\{in\omega_R t\} - 1] \right\} = \sum_{n=-\infty}^{\infty} e_n \exp\{in\omega_R t\}; \tag{118}$$

with an analogous expression for the coefficients d_n^-. Here we have applied two transformations in one step in order to diagonalize the matrices with respect to both the chemical shifts and dipolar terms of H_F^{11} and H_F^{44}. The sequential application of the transformations is valid because the operators Z_n^{pp} and Z_m^{pp} are always commuting. The operators Z_0^{pp} correspond to identity matrices within the blocks p, according to the definition given by Eq. (52).

The submatrix H_F^{23} contains off-diagonal elements that make its diagonalization cumbersome. A similar block-diagonalization of H_F^{23} leads to the expression:

$$(D_F^{22})^{-1}(D_F^{33})^{-1}H_F^{23}(D_F^{33})(D_F^{22}) = \omega_R[N^{22} + N^{33}] - \Delta\omega[Z_0^{22} - Z_0^{33}]$$
$$- \sum_{n=-2}^{2} \{\omega_n^D (D_F^{22})^{-1} P_n^{23} (D_F^{33}) + \omega_n^{D*} (D_F^{33})^{-1} M_{-n}^{23} (D_F^{22})\}, \qquad (119)$$

where:

$$D_F^{22} = 2Z_0^{11} + \left\{\sum_{n=-\infty}^{\infty} 2d_n^- Z_n^{22}\right\}\left\{\sum_{n=-\infty}^{\infty} 2e_{-n}^* Z_n^{22}\right\} + 2Z_0^{33} + 2Z_0^{44};$$

$$D_F^{33} = 2Z_0^{11} + 2Z_0^{22} + \left\{\sum_{n=-\infty}^{\infty} 2d_{-n}^{-*} Z_n^{33}\right\}\left\{\sum_{n=-\infty}^{\infty} 2e_{-n}^* Z_n^{33}\right\} + 2Z_0^{44} \qquad (120)$$

This transformed Floquet Hamiltonian, which is diagonal in all operators Z_n^{pp} with respect to n, still contains off-diagonal elements connecting the states $|2m\rangle$ and $|3m-n\rangle$. A final transformation to complete the diagonalization can be accomplished by numerical methods or perturbation theory [66]. The rigorous result, however, is that the diagonalized Hamiltonian H_F^{23} assumes the form [59]:

$$\Lambda_F^{23} = \omega_R[N^{22} + N^{33}] - (\Delta\omega + \delta\omega)[Z_0^{22} - Z_0^{33}] \qquad (121)$$

4.1.2 Lineshapes at Rotational Resonance

The flip-flop part of the homonuclear dipolar interaction results in a shift $\pm\frac{1}{2}\delta\omega$ of the diagonal elements. The FID signal, which is generated by the transition operator $[M_0^{12} + M_0^{34} + M_0^{13} + M_0^{24}]$, consists of four sets of frequencies which include this shift. The transition operator is the Floquet analogue of the operator $[I_{-1} + I_{-2}]$ representing the observable transverse magnetization. The four types of allowed transition frequencies are the differences between the diagonal elements of the Hamiltonian in Eqs. (116) and (121). The transition amplitudes and frequencies characterizing the MAS spectrum are given by the following expressions [59, 65]:

$$\langle\lambda_n^2|M_0^{12} + M_0^{13}|\lambda_0^1\rangle, \quad -\tfrac{1}{2}\Delta\omega - \tfrac{1}{2}\delta\omega - \tfrac{1}{2}\omega^J + n\omega_R;$$
$$\langle\lambda_n^3|M_0^{12} + M_0^{13}|\lambda_0^1\rangle, \quad \tfrac{1}{2}\Delta\omega + \tfrac{1}{2}\delta\omega - \tfrac{1}{2}\omega^J + n\omega_R;$$
$$\langle\lambda_n^4|M_0^{24} + M_0^{34}|\lambda_0^2\rangle, \quad \tfrac{1}{2}\Delta\omega + \tfrac{1}{2}\delta\omega + \tfrac{1}{2}\omega^J + n\omega_R;$$
$$\langle\lambda_n^4|M_0^{24} + M_0^{34}|\lambda_0^3\rangle, \quad -\tfrac{1}{2}\Delta\omega - \tfrac{1}{2}\delta\omega + \tfrac{1}{2}\omega^J + n\omega_R \qquad (122)$$

Here we have temporarily restored the J-coupling to the homonuclear MAS spectrum in order to illustrate its effect on the line positions. While the effect of the dipolar coupling is to shift the sideband manifolds by $\pm 1/2\,\delta\omega$, the

J-coupling causes splittings of each line. The intensity of each spectral line is a function of the coefficients $d_n^{\pm}$ and e_n, and it also depends on the final transformation used to diagonalize the matrix H_F^{23}. The first and fourth sets of frequencies correspond to the centerband of spin 2 and its sidebands, split by the J-coupling, and likewise the second and third sideband manifolds correspond to the MAS spectrum of spin 1.

The homonuclear dipolar interaction shifts these four sets of spectral lines, but does not create additional ones [59]. In other treatments [41, 122], eight sets of frequencies arise from a dipolar splitting induced at rotational resonance, in contrast to the frequency shift discussed here. Eight sets of sideband manifolds arise in the context of an interaction frame defined by the chemical shift evolution of the two spins [122]. However, each manifold has a partner with which it is perfectly overlapping in the frequency domain, so it is possible to recombine the eight manifolds into four unique ones characterized by a shift from the dipolar coupling and a splitting from the J-coupling.

If the dipolar interaction is smaller than the spinning frequency, $\omega_n^D \ll \omega_R$, then the off-diagonal elements in Eq. (119) can usually be neglected. Under these circumstances, the Hamiltonian resembles that of the heteronuclear case, and no shifts $1/2\,\delta\omega$ are observed. However, when the off-diagonal terms connect nearly degenerate diagonal elements, even small flip-flop terms influence the eigenvalues of the Floquet Hamiltonian, and $1/2\,\delta\omega$ becomes significant. The shift $1/2\,\delta\omega$ lifts the degeneracy that occurs at the rotational resonance condition, $\Delta\omega \approx N\omega_R$, and leads to avoided crossings of the Floquet states, as illustrated in Fig. 23 [59].

Near rotational resonance, only the terms in Eq. (119) involving $P_N^{23} = X_N^{23} + iY_N^{23}$ and $M_{-N}^{23} = X_{-N}^{23} - iY_{-N}^{23}$ must be included, and their coefficients take the form:

$$\bar{\omega}_{\text{eff}}(N) = \sum_{m=-\infty}^{\infty} \sum_{n=-2}^{2} \{-d_{m-N}^{-*}\omega_n^D d_{n-m}^{-*}\}, \tag{123}$$

again omitting the J-coupling. Equation (123) results from the application of the transformation $D_F^{22} D_F^{33}$ to the terms involving M_n^{23} and P_n^{23}, as expressed by Eq. (119), followed by the elimination of all non-resonant terms. The relevant part of the Z-diagonalized Hamiltonian H_F^{23} can then be written in the form:

$$H_F'^{23} \approx \omega_R[N^{22} + N^{33}] - \Delta\omega[Z_0^{22} - Z_0^{33}] + \bar{\omega}_{\text{eff}}^D(N) P_N^{23} + \bar{\omega}_{\text{eff}}^{D*}(N) M_{-N}^{23} \tag{124}$$

At exact rotational resonance, $\Delta\omega = N\omega_R$, the Hamiltonian of Eq. (124) is converted into the form of Eq. (121) by 2×2 diagonalizations, and the homonuclear frequency shift $1/2\,\delta\omega$ assumes the simple form:

$$\tfrac{1}{2}\delta\omega = |\bar{\omega}_{\text{eff}}^D(N)| \tag{125}$$

This derivation is similar to the Floquet analysis of rotary resonance [37], which involved parallel diagonalizations of the two related submatrices shown

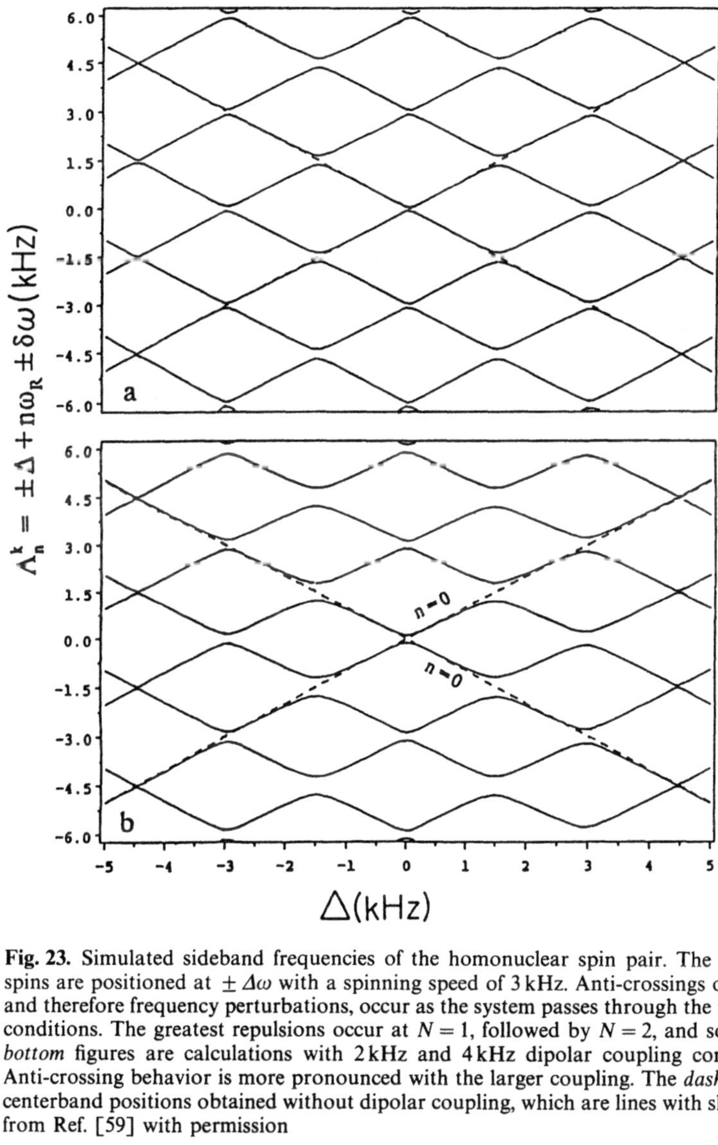

Fig. 23. Simulated sideband frequencies of the homonuclear spin pair. The spectral lines of the spins are positioned at $\pm \Delta\omega$ with a spinning speed of 3 kHz. Anti-crossings of the Floquet states, and therefore frequency perturbations, occur as the system passes through the rotational resonance conditions. The greatest repulsions occur at $N = 1$, followed by $N = 2$, and so forth. The *top* and *bottom* figures are calculations with 2 kHz and 4 kHz dipolar coupling constants, respectively. Anti-crossing behavior is more pronounced with the larger coupling. The *dashed lines* indicate the centerband positions obtained without dipolar coupling, which are lines with slopes ± 1. Reprinted from Ref. [59] with permission

in Fig. 6. In this case, however, we show how the off-diagonal terms are modified by differences in CSA between the spins, according to Eq. (123). The 2×2 transformations involving pairs of nearly degenerate Floquet states provide a straightforward approach to understanding resonance behavior in MAS experiments.

For CSA parameters that are much smaller than the spinning speed, the coefficients d_n^- are negligible except for $d_0^- \approx 1$. In this case, $|\bar{\omega}_{\text{eff}}^{\text{D}}(N)| = |\omega_{\text{N}}^{\text{D}}|$ at the rotational resonance conditions $N = \pm 1, \pm 2$, and $\delta\omega \approx 0$ for $|N| > 2$. In powder samples, each crystallite evolves according to its own shift, which depends on ω_N^D and hence the orientational angles (θ, ϕ) of the internuclear vector. Once again, the distribution of crystallites in a powder sample leads to inhomogeneous powder lineshapes. These lineshapes can be calculated by numerical methods. An exact simulation of the lineshapes of zinc acetate at $N = 1$ rotational resonance is shown in Fig. 24 and compared with the experimental spectrum near the centerbands [41]. The contributions from several crystallites are shown with the powder average. In contrast to the case of MW inhomogeneous spin systems, Fig. 24c demonstrates that an average over the rotor phase angle α of the crystallites does not lead to a uniform phase across the spectrum.

To summarize, the inhomogeneous shifts $1/2\,\delta\omega$ are caused by the perturbation of degeneracies between Floquet states that occurs at rotational resonance. For $N = \pm 1, \pm 2$, in the limit of relatively small CSA tensors, the shift $|\bar{\omega}_{\mathrm{eff}}^{\mathrm{D}}(N)|$ is approximately equal to the magnitude of the dipolar Fourier

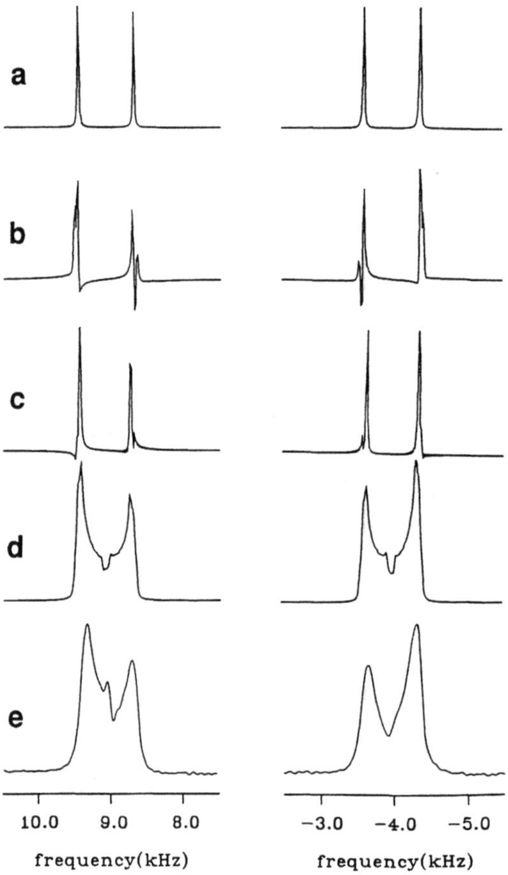

Fig. 24. Calculated rotational resonance lineshapes of zinc acetate in the region near the centerbands. (*a–b*) Two single crystal MAS spectra are shown with Euler angles $(0, \pi/4, 0)$ and $(\pi/2, \pi/4, 0)$, which relate the molecular reference frame (chosen to be the carboxyl CSA tensor) to the rotor frame; (*c*) an average over crystallite orientations with $(\alpha, \pi/4, 0)$, which does not yield an absorptive spectrum; (*d*) a complete powder average; (*e*) comparison with experimental results. Reprinted from Ref. [41] with permission

component $|\omega_N^D|$ that is resonant with the sample rotation. In general, significant resonances also occur for $|N| > 2$, with shifts that involve a more complicated convolution of the dipolar interaction with the CSA interactions of the two spins, including their orientations with respect to each other and the principal axis system of the dipole-dipole interaction, which is defined by the internuclear vector. To calculate lineshapes accurately at rotational resonance, it is necessary to estimate the relationships among the principal axis systems of the three anisotropic spin interactions.

4.1.3 Magnetization Exchange Experiments

For the relatively long internuclear distances that are of the greatest interest in structural investigations, $r > 3$Å, lineshape perturbations at rotational resonance are weak and difficult to observe. Therefore, it is desirable to measure internuclear distances by replacing the lineshape effect, whose signal decays with time constant T_2^*, with an experiment that permits observation of the spin system on a much longer time scale. The flip-flop interaction that is selectively reintroduced into MAS experiments at rotational resonance can be used to drive longitudinal exchange, which is observable on the time scale of T_1. The influence of weak couplings on magnetization exchange trajectories is often observable at rotational resonance, even when their effects on the lineshapes are too small to observe. Figure 25 demonstrates the pulse sequence used to prepare difference magnetization between two spins and monitor its time-evolution. At $N = 3$ rotational resonance, Fig. 26 illustrates the trajectory of magnetization exchange in doubly ^{13}C-labeled zinc acetate following selective inversion of one of the resonances. Selective inversion prepares the initial state of difference polarization $\rho(0) = 1/2\,[I_{z1} - I_{z2}]$ [71, 129]. At rotational resonance, the effective dipolar flip-flop interaction greatly enhances the rate of exchange among homonuclear spins whose relative longitudinal polarizations lie away from equilibrium.

This effect was observed initially by Andrew et al. [121] in ^{31}P spectra of PCl_5 (consisting of PCl_4^+ and PCl_6^-). In particular, they observed that the longitudinal relaxation times of the distinguishable ^{31}P resonances, which normally differ from one another by a factor of ten, become equal at rotational resonance, suggesting that the mechanical motion of the rotor drives spin exchange. More recently, in inversion exchange experiments, enhanced rates of

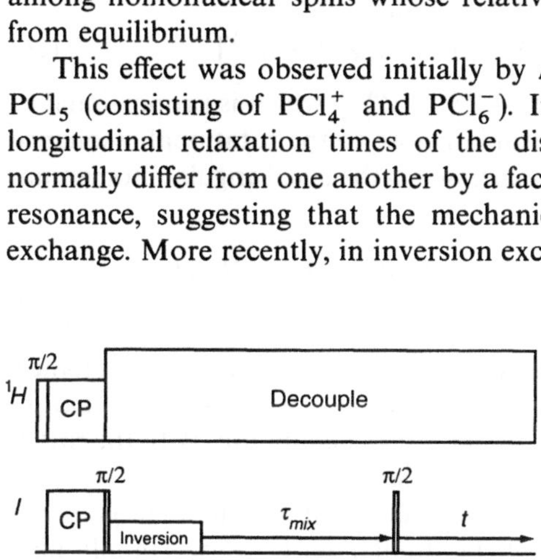

Fig. 25. Pulse Sequence for Magnetization Exchange. The spins are polarized and stored along the magnetic field direction. Following selective inversion of one of the resonances, exchange occurs at rotational resonance, and the signal is acquired

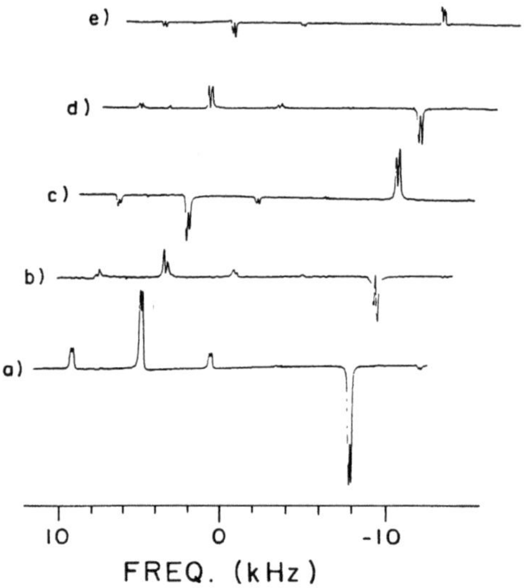

Fig. 26. Magnetization Exchange Experiment. (*a–e*) Exchange is demonstrated in doubly-^{13}C labeled zinc acetate at $N = 3$ rotational resonance, whose spectra are shown after 0.0, 1.0, 3.0, 5.5, and 7.0 ms of mixing, respectively. Reprinted from Ref. [40] with permission

exchange have been observed in doubly ^{13}C-labeled samples and used to measure ^{13}C-^{13}C distances [40, 41]. Such spin pairs can often be engineered by selective isotopic enrichment [130]. The enhancement of spin diffusion among natural abundance ^{13}C spins has also been observed [42–44].

In order to analyze longitudinal exchange experiments, we derive an effective homonuclear Hamiltonian that is valid for synchronous sampling. The approximate result of Eq. (124) can be adapted for this purpose. With a suitable rearrangement of the matrix indices, the diagonalized Floquet Hamiltonian can be written as follows:

$$H_F'^{23} = \omega_R[N^{22} + N^{33}] - (\Delta\omega - N\omega_R)[Z_0^{22} - Z_0^{33}] \\ + \bar{\omega}_{\text{eff}}^{D}(N)P_0^{23} + \bar{\omega}_{\text{eff}}^{D*}(N)M_0^{23} + N\omega_R[Z_0^{22} + Z_0^{33}] \tag{126}$$

After subtraction of $1/2\, N\omega_R[Z_0^{11} + Z_0^{22} + Z_0^{33} + Z_0^{44}]$, which is proportional to the identity matrix, the effective Hamiltonian close to the rotational resonance condition $\Delta\omega \approx N\omega_R$ assumes the form, expressed in the usual angular momentum operators:

$$\bar{H}_{\text{eff}} = \tfrac{1}{2}(\Delta\omega - N\omega_R)[I_{z1} - I_{z2}] + |\bar{\omega}_{\text{eff}}^{D}(N)|[I_{+1}I_{-2} + I_{-1}I_{+2}] \\ - N\omega_R I_{z1}I_{z2} \tag{127}$$

The effective Hamiltonian consists of an offset from exact rotational resonance, an homonuclear flip-flop term, and a term $-N\omega_R I_{z1}I_{z2}$, which does not participate in the longitudinal exchange experiment, since it commutes with the rest of the Hamiltonian as well as all relevant spin coherences.

At exact rotational resonance, the difference polarization undergoes the following trajectory under the action of the dipolar term in the Hamiltonian of Eq. (127):

$$\rho(M\tau_R) = \tfrac{1}{2}[I_{z1} - I_{z2}]\cos(2|\bar{\omega}^D_{eff}(N)|M\tau_R) + [I_{x1}I_{y2} - I_{y1}I_{x2}]\sin(2|\bar{\omega}^D_{eff}(N)|M\tau_R) \quad (128)$$

Any component of sum polarization $1/2\,[I_{z1} + I_{z2}]$, on the other hand, is a constant of the motion. Within each crystallite, the difference polarization $1/2\,[I_{z1} - I_{z2}]$ exchanges with the zero-quantum coherence, which has the form $[I_{x1}I_{y2} - I_{y1}I_{x2}]$, in an oscillatory fashion. Since each crystallite oscillates with a somewhat different frequency, the observed signal from a powder sample decays inhomogeneously [41, 131] as in the case of the FID. The oscillatory decay observed with a relatively strong dipolar coupling is shown in Fig. 27, again for the case of zinc acetate. The quenching of the exchange effect that occurs away from rotational resonance is also demonstrated. The flip-flop interaction can also be used for other purposes, such as double-quantum filtering [132].

So far, we have neglected relaxation. However, because the two terms of the density matrix in Eq. (128) generally relax at much different rates, it is necessary to consider the finite damping rate of the zero-quantum coherence, which decays on the time scale of T_2. On the other hand, the longitudinal magnetization of a dilute spin decays much more slowly, on the order of T_1, and so its relaxation process can be safely neglected. This situation contrasts with the case of dephasing experiments, where all of the relevant terms of the density matrix decay at comparable rates. For example, in REDOR experiments, the dephasing of transverse magnetization S_x occurs through exchange with

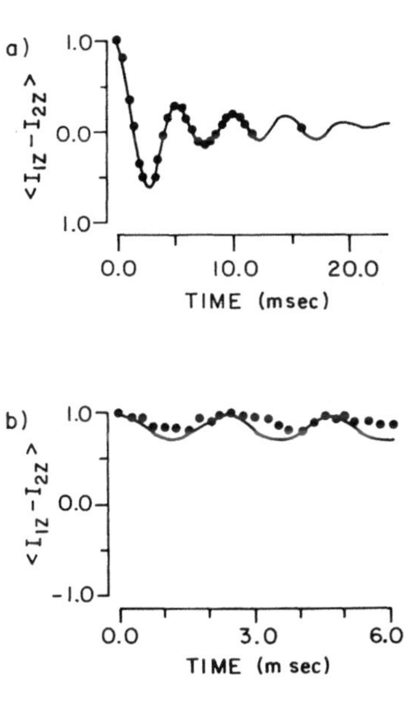

Fig. 27a, b. Magnetization Exchange Trajectories. The time evolution of the polarization difference is plotted for the same zinc acetate sample and compared with numerical simulations. The results at two spinning speeds are shown: **a** 4.378 kHz, **b** 4.26 kHz, which corresponds to a 120 Hz deviation from rotational resonance. Reprinted from Ref. [40] with permission single quantum coherence of the form $2S_yI_z$. Under the reasonable assumption that all terms decay at comparable rates, the relaxation dynamics are essentially decoupled from the dynamics of dipolar dephasing. Thus, only the latter is reflected in the REDOR difference signal. In longitudinal exchange experiments on homonuclear spin pairs, the influence of relaxation becomes important when the decay constant of the zero-quantum coherence is comparable to the size of the dipolar coupling. This situation often arises in experiments where weak dipolar couplings are examined, which correspond to relatively long internuclear distances.

One important case in which zero-quantum damping dominates the exchange dynamics is spin diffusion among dilute spins in the absence of the proton decoupling [42–44]. In this situation, the resonances of the dilute spins acquire broad homogeneous linewidths, which arise from coupling to the homogeneous network of coupled protons, and the damping of all off-diagonal coherences is fast compared to the dipolar couplings among dilute spins. Kubo and McDowell have derived expressions for the rate of magnetization transfer in this limit [43]. They also provided an approximate expression for the relaxation time of the zero-quantum coherence $[I_{x1}I_{y2} - I_{y1}I_{x2}]$ in terms of single spin tranverse relaxation times:

$$\frac{1}{T_2^{ZQ}} \approx \frac{1}{T_2^1} + \frac{1}{T_2^2}, \tag{129}$$

where T_2^i is the homogeneous transverse relaxation time of spin i and T_2^{ZQ} is the relaxation time of the zero-quantum coherence of the spin pair. Equation (129) is valid when the rapidly fluctuating interactions promoting the process of homogeneous decay are uncorrelated between the two spins [43]. Even with the use of proton decoupling, zero-quantum dephasing, which arises from residual contributions to the homogeneous linewidths of the two resonances, must be included in the analysis of the trajectories when it is fast enough to compete with the dynamical evolution induced by the dipolar flip-flop term itself. Therefore, in the case of distant spins with weak dipolar couplings, zero-quantum relaxation plays an important role in the exchange dynamics. In addition, the zero-quantum damping rate is a useful phenomenological parameter in describing other effects, such as the departures from exact rotational resonance within an inhomogeneous linewidth and instabilities in the frequency of rotation.

Analytical expressions describing the behavior of the difference polarization at exact rotational resonance have been derived in limiting cases [41]. In the limit of slow dephasing, the trajectory for each crystallite is a damped oscillation:

$$\langle I_{z1} - I_{z2} \rangle(t) \approx \exp\{-t/2T_2^{ZQ}\} \cos(2|\bar{\omega}_{\text{eff}}^{D}(N)|t) \tag{130}$$

On the other hand, when the dephasing is rapid compared to the flip-flop coupling, the difference polarization of each crystallite decays exponentially as follows:

$$\langle I_{z1} - I_{z2} \rangle(t) \approx \exp\{-4|\bar{\omega}_{\text{eff}}^{D}(N)|^2 T_2^{ZQ} t\} \tag{131}$$

As the rate of zero-quantum dephasing accelerates, T_2^{ZQ} becomes shorter, and according to Eq. (131) the difference magnetization between the two spins pair decays more slowly [41]. Therefore, suppression of the dephasing process with strong proton decoupling enhances the rate of magnetization transfer and makes it easier to observe the impact of the dipolar coupling on the exchange trajectory.

With an estimate of the homogeneous linewidths of the two resonances and their dipolar and chemical tensors, including all relative orientations, the magnetization exchange trajectory of each crystallite can be calculated with exact or approximate numerical simulations [41]. Altogether, five angles of relative orientation are required in order to define the relationships among the dipolar vector and the CSA tensor directions. The overall powder signal from a spin pair can be obtained by averaging the results together from a large number of crystallite orientations. Comparison of simulated trajectories with experimental results leads to a measurement of the dipolar coupling constant and hence to the internuclear distance between the spins. Several applications of this approach to the measurement of homonuclear dipolar couplings have been reported [133–138].

4.2 Multiple Pulse Recoupling

4.2.1 Solid Echo Recoupling

In many cases, it is not practical to meet the rotational resonance condition. For example, if the two spins have nearly degenerate isotropic chemical shifts, then very slow spinning becomes necessary to induce dipolar recoupling. However, as in the case of heteronuclear coupled spin pairs, synchronously applied pulses can significantly enhance the dipolar decay of transverse coherence [45]. In addition, multiple quantum transitions can be generated in coupled homonuclear spin systems [46].

Based on the transformation properties of spherical tensor components, an interesting set of pulse sequences was derived by Tycko to generate spectra resembling those that would be obtained in zero magnetic field [139, 140]. These experiments have succeeded in revealing the zero-field dipolar spectra of rotating spin pairs in a high magnetic field. Tycko and Dabbagh [45] have also introduced the Dipolar Recovery at the Magic Angle (DRAMA) experiment, which enables the observation of dipolar oscillations in the rotational echo amplitudes of homonuclear spin pairs. The experiment is based upon the application of two $\pi/2$ pulses per rotor cycle and the acquisition of the signal at rotational echo positions. In effect, these sequences generate partial solid echoes [10] in spin space that operate contrary to the modulation imposed by MAS and, as a consequence, lead to direct dipolar recoupling. In this context, "direct" recoupling indicates that no involvement of the chemical shift interactions is essential to the process of restoring the dipole-dipole coupling to the MAS experiment.

The DRAMA scheme is most efficient close to the $N = 0$ rotational resonance condition, where the two spins have similar isotropic chemical shifts. It is precisely under these circumstances that magnetization exchange experiments at rotational resonance become impractical. To diminish the influence of the CS parameters on the dipolar decay in DRAMA experiments, two π pulses are included in the basic DRAMA cycle. Thus, the basic pulse cycle, which is shown in Fig. 28, extends over four rotor periods and consists of ten pulses. To calculate the evolution of nuclear spins under the influence of this pulse sequence, it is most convenient to calculate the effective Hamiltonian in the toggling frame using the AHT approach, following the treatment of Tycko et al. [45].

To do so, using the definitions in Eq. (114) and the usual spin angular momentum operators, we rewrite the spin Hamiltonian in the form:

$$H(t) = \omega^{\mathrm{D}}(t)H_{zz} - \tfrac{1}{2}(\Delta\omega + \omega^{-}(t))[I_{z1} - I_{z2}] - \tfrac{1}{2}\omega^{+}(t)[I_{z1} + I_{z2}], \quad (132)$$

where H_{zz} is one of the bilinear operators H_{aa} defined by:

$$H_{\mathrm{aa}} = [3I_{\mathrm{a1}}I_{\mathrm{a2}} - \vec{I}_1 \cdot \vec{I}_2], \quad (133)$$

and $a = x, y, z$. The $\omega^{\pm}(t)$ parameters are again the sums and differences of the CSA coefficients of two spins, and the J-coupling is neglected for convenience. To evaluate the average Hamiltonian, $H(t)$ is transformed into the toggling frame. All pulses in the DRAMA experiment are applied in the X direction in the rotating frame and therefore do not affect H_{xx}. Each $\pi/2$ pulse inverts the operator $[H_{zz} - H_{\mathrm{yy}}]$, but leaves the terms involving $[H_{zz} + H_{\mathrm{yy}}]$ unchanged. The $\pi/2$ inversion of $[H_{zz} - H_{\mathrm{yy}}]$ is similar to the inversion of I_z by π pulses, and so the basic DRAMA experiment can be viewed as an homonuclear analogue of the REDOR experiment. On the other hand, the π pulses do not alter the H_{aa} operators. In the toggling frame, the complete spin Hamiltonian can be expressed:

$$\begin{aligned} H^{\mathrm{tg}}(t) = {} & \tfrac{1}{2}\omega^{\mathrm{D}}(t)P(t)[H_{zz} - H_{\mathrm{yy}}] + \tfrac{1}{2}\omega^{\mathrm{D}}(t)[H_{zz} + H_{\mathrm{yy}}] \\ & - \tfrac{1}{2}(\Delta\omega + \omega^{-}(t))\{Q(t)[I_{z1} - I_{z2}] + R(t)[I_{\mathrm{y1}} - I_{\mathrm{y2}}]\} \\ & - \tfrac{1}{2}\omega^{+}(t)\{Q(t)[I_{z1} + I_{z2}] + R(t)[I_{\mathrm{y1}} + I_{\mathrm{y2}}]\}, \end{aligned} \quad (134)$$

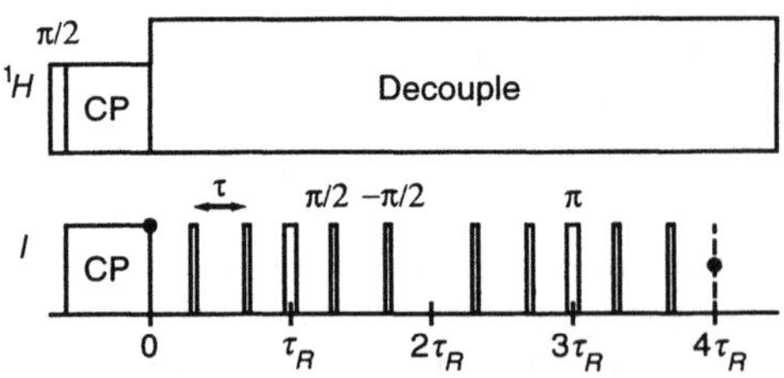

Fig. 28. Pulse sequence for the DRAMA experiment. All pulses are applied with phases X or $\bar{\mathrm{X}}$ in the rotating frame. Each pair of $\pi/2$ pulses has alternating phase X, $\bar{\mathrm{X}}$, and the π pulses are also alternated by X, $\bar{\mathrm{X}}$, which leads to toggling frame functions that are symmetric in time. The $\pi/2$ pulses spoil dipolar refocusing by MAS

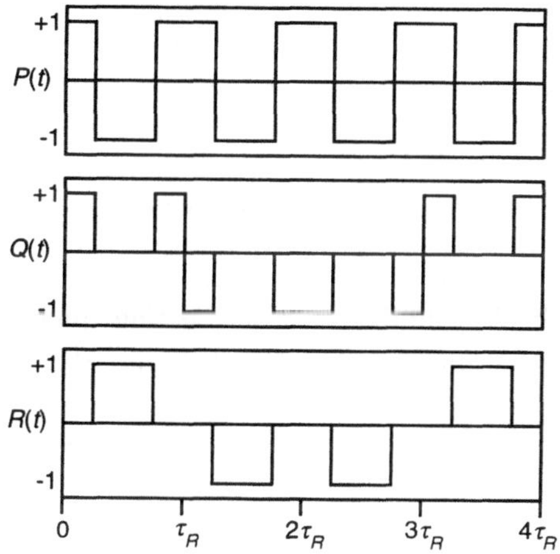

Fig. 29. Toggling frame functions for DRAMA. These functions characterize the influence of the pulses on the internal spin Hamiltonian

where the three functions $P(t), Q(t)$, and $R(t)$, sketched in Fig. 29, define the action of the RF pulses in the toggling frame for the particular case: $\tau = 1/2\,\tau_R$.

To evaluate the average Hamiltonian, it is again most convenient to utilize the Fourier transforms of the toggling functions. Because the length of the DRAMA cycle is $t_C = 4\tau_R$, the basic frequency unit of the transform is $1/4\,\omega_R$. In these units, the dipolar and CS Fourier components for $n \neq 0$ have indices equal to $\pm 4, \pm 8$ in place of $\pm 1, \pm 2$. However, if the earlier definitions of the coefficients ω_n^D, ω_n^+, and ω_n^- are retained, then the following effective Hamiltonian is obtained from the first term of the Magnus expansion:

$$
\begin{aligned}
\bar{H}_{\text{eff}}^{(0)} &= \frac{1}{4\tau_R} \int_0^{4\tau_R} H^{\text{tg}}(t)dt \\
&= \tfrac{1}{2}(2\omega_1^D P_{-4} + 2\omega_{-1}^D P_4)[H_{zz} - H_{yy}] \\
&= \bar{\omega}_{\text{eff}}^D [H_{zz} - H_{yy}],
\end{aligned}
\tag{135}
$$

where the effective dipolar frequency recovered by the pulse sequence is given by:

$$
\bar{\omega}_{\text{eff}}^D = \frac{\sqrt{2}}{2\pi}\,\omega_{12}^D \sin 2\theta \cos\phi \tag{136}
$$

In this approximation, all CS terms are averaged to zero, and the average Hamiltonian of the DRAMA experiment depends only on the interaction between the two spins and their orientation (θ, ϕ).

Neglecting higher order terms in the Magnus expansion, the transverse magnetization, when sampled synchronously with the pulse cycle, behaves as follows:

$$
\rho(t) = [I_{x1} + I_{x2}]\cos(2\bar{\omega}_{\text{eff}}^D t) + 2[I_{y1}I_{z2} + I_{z1}I_{y2}]\sin(2\omega_{\text{eff}}^D t) \tag{137}
$$

Integration of the amplitude of the transverse magnetization $\cos(2\bar{\omega}^{D}_{eff}t)$ over all angles (θ, ϕ) results in a decay trajectory whose Fourier transform is a powder pattern [45]. Several powder lineshapes, corresponding to various ways of positioning the $\pi/2$ pulses within the rotor cycle, are illustrated in Fig. 30. The bilinear coherence $[I_{y1}I_{z2} + I_{z1}I_{y2}]$ cannot be observed directly, but it is useful as a pathway to differentiate coupled from uncoupled spins. Tycko and Smith [46] have shown that the DRAMA sequence can be applied as a double quantum filter to excite only coupled spin pairs and generate MAS spectra that do not exhibit spectral lines belonging to the background of non-interacting nuclei. This approach can be used to simplify MAS spectra or to detect the proximity of nuclear spins.

The AHT analysis shows how the DRAMA experiment recovers a non-vanishing effective dipolar coupling from the MAS Hamiltonian. The pulse sequence renders the spin operator portion of the dipolar coupling time-dependent in the toggling frame, where modulations arise that are characterized by the toggling functions $P(t)$, $Q(t)$, and $R(t)$. This time-dependence, in turn, destructively interferes with the amplitude modulation of the dipolar coupling by magic angle spinning. The consequence of the combined modulations imposed by MAS and the RF pulse sequence is a non-zero coupling averaged over time. This interference occurs when the toggling functions have Fourier components matching at least one of the frequency modes of the MAS amplitude modulation. The latter has Fourier components only through $|m| = 2$.

In a similar way, an effective heteronuclear interaction is recovered in REDOR experiments, where π pulse sequences are used to cause interference between the spin and spatial factors of the dipolar coupling in the toggling

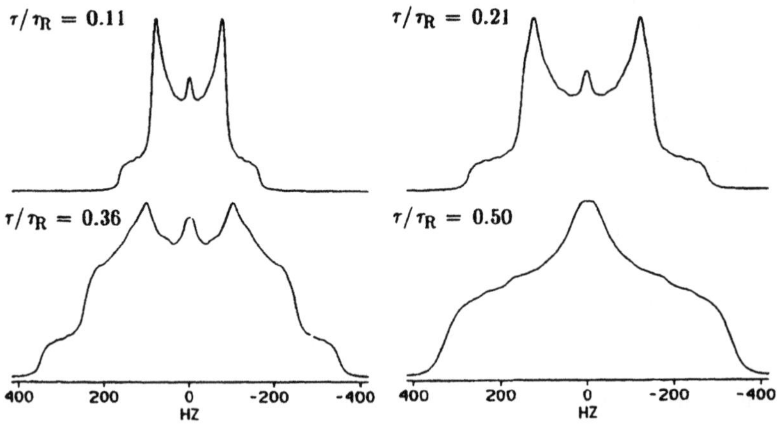

Fig. 30. Calculated Fourier transforms of DRAMA dephasing trajectories. Dipolar powder spectra obtained with various pulse spacings τ are shown. The coupling constant is 1.5 kHz, the rotor period 399 μs, and the π pulse length 10 μs. Reprinted from Ref. [45] with permission

frame. However, in the case of REDOR, the average Hamiltonian can be calculated *exactly* using AHT because of the commutation of all components of the toggling frame Hamiltonian at different times. This property is a very useful one in developing experiments to study heteronuclear spin systems, but unfortunately it is not applicable to homonuclear spin systems.

Since self-commutation in the toggling frame does not hold in DRAMA experiments, the zeroth order result is valid only for interactions that are small relative to the cycle time $4\tau_R$. This restriction is usually obeyed for the dipolar coupling constants ω_{12}^D found in dilute spin systems. However, in many experiments, the terms of the CS interaction can be at least as large as the DRAMA cycle frequency $1/4\omega_R$. Under these conditions, the DRAMA decay becomes a strong function of the CS parameters, and information about these parameters becomes necessary to evaluate the strength of the dipolar interaction. Extended versions of the basic DRAMA sequence have been developed in order to minimize these effects [46].

Direct recoupling of the dipolar interaction can also be achieved by the Unified Spin Echo and Magic Echo (USEME) approach of Fujiwara et al. [141, 142], in which a continuous RF field is applied during each second half rotor cycle to scale the homonuclear dipole-dipole coupling by $-1/2$. In static solids, since the sign of the interaction is flipped, the application of an RF field can be used to reverse the time-evolution of the homonuclear spin system and form "magic echoes" [143]. In rotating solids, however, sign reversal during a portion of the rotor period leads to the cancellation of rotational refocusing. In a similar way, the application of phase-alternated $\pi/2$ pulses in DRAMA leads to the cancellation of the dipolar contribution $1/2\,[H_{zz} - H_{yy}]$ in the static case, but to the recoupling of the same interaction in the spinning case. The usual form of the dipolar spin operators is preserved during the entire USEME cycle, so the dipolar coupling is self-commuting during the two time periods in the sequence. The zeroth order result therefore yields the exact effective Hamiltonian in the absence of chemical shifts.

4.2.2 *Spin Echo Techniques*

Another approach for the enhancement of the dipolar dephasing of homonuclear spin pairs was introduced by Gullion and Vega [47], who applied a single π pulse during each rotor period and detected the signal synchronously at the rotational echo positions. The result of this Simple Excitation for the Dephasing of the Rotational Echo Amplitudes (SEDRA) is again dependent on the CSA parameters of the spin system, but mainly on the difference between the isotropic chemical shifts of the two spins $\Delta\omega$. In Fig. 31, the basic SEDRA pulse sequence is shown. The basic element consists of two rotor periods with π pulses applied in the middle of each period, which fully refocus anisotropic, as well as isotropic, chemical shifts [51]. Consequently, signals from uncoupled spins are fully refocused after each pair of pulses.

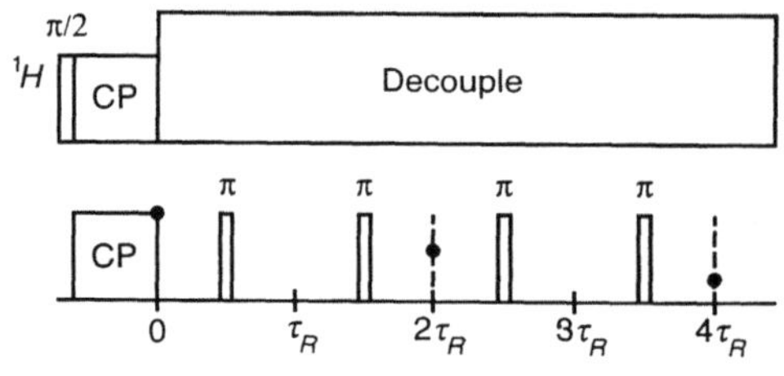

Fig. 31. Pulse Sequence for the SEDRA dephasing experiment. Dephasing of the magnetization is observed with one π pulse per rotor period

Likewise, if two coupled spins have identical CS interactions, then their dipolar coupling, which is not affected by π pulse sequences, behaves inhomogeneously in the MW sense, and observed signals again refocus completely after multiples of the rotor period [22, 47, 48]. However, in the more general case where there are significant differences in chemical shifts between the spins, the dipolar interaction is partially recoupled into the MAS experiment by rotor-synchronized π pulses. The mechanism of recoupling is based upon the modulation of the MAS dipolar coupling by the chemical shift differences, as in the case of rotational resonance. However, because of the refocusing of all CS terms by the spin echo sequence, dipolar recoupling occurs over a much broader range of frequencies.

Floquet Theory and Average Hamiltonian Theory have been applied to evaluate an effective Hamiltonian that is suitable for the description of homonuclear recoupling experiments involving spin echo sequences [47, 48]. In both approaches, the toggling frame transformation is essential to the derivation. After conversion into the toggling frame, the CS terms assume the form:

$$H_{\mathrm{CS}}^{\mathrm{tg}}(t) = -(\Delta\omega + \omega^{-}(t))F(t)[I_z^{22} - I_z^{33}] - \omega^{+}(t)F(t)[I_z^{11} - I_z^{44}] \qquad (138)$$

The toggling frame function $F(t)$, illustrated in Fig. 32, describes the effect of the RF pulses, and switches between $+1$ and -1 under the assumption that the π pulses are ideal rotations. The CS difference term $\Delta\omega$ now becomes time-dependent through the modulation described by $F(t)$, and the pulses also modify the time-dependence of the CSA terms. In contrast to the heteronuclear case, the dipolar coupling is invariant with respect to the toggling frame transformation defined by the π pulses, since the pulses simultaneously flip the signs of the angular momentum operators of both spins. Accordingly, the dipolar coupling retains its original form in the toggling frame:

$$H_{\mathrm{D}}^{\mathrm{tg}}(t) = \omega^{\mathrm{D}}(t)[I_z^{11} - I_z^{22} - I_z^{33} + I_z^{44}] - \omega^{\mathrm{D}}(t)I_x^{23} \qquad (139)$$

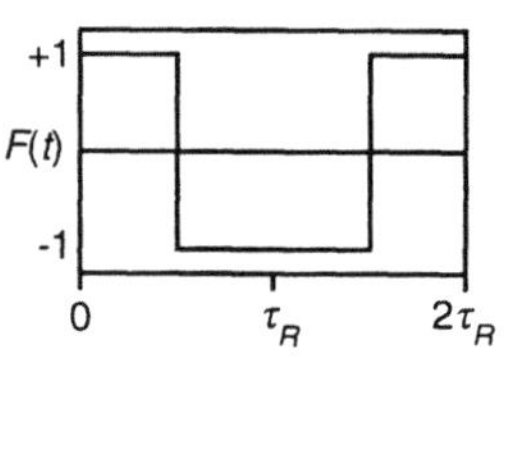

Fig. 32. Toggling frame function. Square wave modulation is superimposed on all chemical shift terms in the toggling frame

The basic SEDRA cycle has a length of $t_C = 2\tau_R$ over which it is cyclic. In the Floquet approach, the Hamiltonian is written in terms of the Fourier components of $\omega^{\pm}(t)$, $\omega^{D}(t)$, and $F(t)$, which are expanded as follows:

$$\omega^{D}(t) = \sum_{n=-4}^{4} 2\tilde{\omega}_n^{D} \exp\{in\tfrac{1}{2}\omega_R t\};$$

$$F(t) = \sum_{n=-\infty}^{\infty} F_n \exp\{in\tfrac{1}{2}\omega_R t\}; \tag{140}$$

and likewise for $\omega^{\pm}(t)$. The coefficients $\tilde{\omega}_{2n}^{D}$ are equal to the constants ω_n^{D} that were defined earlier, and the most significant Fourier coefficients of the toggling function $F(t)$ are $F_{\pm 1} = 2/\pi$ and $F_{\pm 3} = 2/3\pi$. Neglecting the CSA terms for simplicity, the Floquet Hamiltonian can be constructed and diagonalized straightforwardly to obtain the effective Hamiltonian. The Floquet operators Z_n^{pp} corresponding to the spin operators I_z^{pp} of Eqs. (138) and (139) can be diagonalized using Eq. (64), where the coefficients d_n of the transformation matrices are defined by the usual relationship:

$$\exp\left\{-\sum_{n=-\infty}^{\infty} \frac{(-\frac{1}{2}\Delta\omega F_n)}{(\frac{1}{2}n\omega_R)}[\exp\{in\tfrac{1}{2}\omega_R t\} - 1]\right\}$$

$$= \sum_{n=-\infty}^{\infty} d_n \exp\{in\tfrac{1}{2}\omega_R t\} \tag{141}$$

For dipolar coupling constants that are small relative to the spinning frequency ω_R, the approximate Floquet Hamiltonian for synchronous sampling takes the form:

$$H_F' = \omega_R[N^{11} + N^{22} + N^{33} + N^{44}] + 2\bar{\omega}_{eff}^{D} X_0^{23}, \tag{142}$$

where:

$$\bar{\omega}_{eff}^{D} = \sum_{m=-\infty}^{\infty} \sum_{k=-4}^{4} \{-d_m^{*}\tilde{\omega}_k^{D} d_{k-m}^{*}\} \tag{143}$$

This Floquet Hamiltonian corresponds to the following effective Hamiltonian expressed in terms of the usual spin operators, which is a pure flip-flop interaction:

$$\bar{H}_{eff} = \bar{\omega}_{eff}^{D}[I_{+1}I_{-2} + I_{-1}I_{+2}] \tag{144}$$

If the J-coupling were included in the analysis, then an additional term $\omega^{J} I_{z1} I_{z2}$ would appear in the effective Hamiltonian for the pulse cycle.

The effective dipolar frequency $\bar{\omega}_{eff}^{D}$ is a function of the Fourier coupling parameters $\tilde{\omega}_k^{D}$ as well as the d_n coefficients. In the case where $\Delta\omega = 0$, it follows from Eq. (140) that only d_0 is non-zero and thus $\bar{\omega}_{eff}^{D} = 0$. On the other hand, for large values of $\Delta\omega$, the d_n values for large n become increasingly significant. Since the coefficients are normalized, $\sum_n d_n = 1$, and it follows that $\bar{\omega}_{eff}^{D}$ is attenuated for large $\Delta\omega$. The maximal dipolar frequency is expected when the chemical shift difference $\Delta\omega$ between the two spins lies in the region between ω_R and $2\omega_R$. This behavior is illustrated in Fig. 33, which provides a plot of the magnitude of the recoupled interaction as a function of spinning speed for two artificial crystallite orientations [48]. The plot illustrates that the maximum recoupling effect is obtained at the $N = 1$ rotational resonance condition for the $m = 1$ component of the MAS dipolar coupling and similarly for the $m = 2$ component. However, recoupling also occurs to a lesser extent at larger and smaller spinning speeds.

This behavior differs fundamentally from that of heteronuclear spin pairs. The CS terms play no role in the dynamics of dipolar recoupling in the heteronuclear case. However, since π pulses have no direct effect on the homonuclear interaction, dipolar recoupling in the homonuclear case arises from interference between the modulation of the flip-flop term by relatively large chemical shift differences and the modulation imposed on the amplitude of the coupling by MAS. In this sense, the recoupling effect generated by spin echo sequences is similar to that of rotational resonance. In contrast, $\pi/2$ pulses are employed in DRAMA to operate directly on the dipolar spin operators in the toggling frame. Dipolar recoupling is thereby generated by interference between the RF modulation of the dipolar spin operators and the oscillations directly imposed upon them by MAS. Although the CS terms must also be considered in the analysis of DRAMA sequences, their influence is not essential to the basic physics of the recoupling process.

The effective dipolar Hamiltonian governs the synchronously detected signal and results in a function that characterizes the decay of transverse magnetization

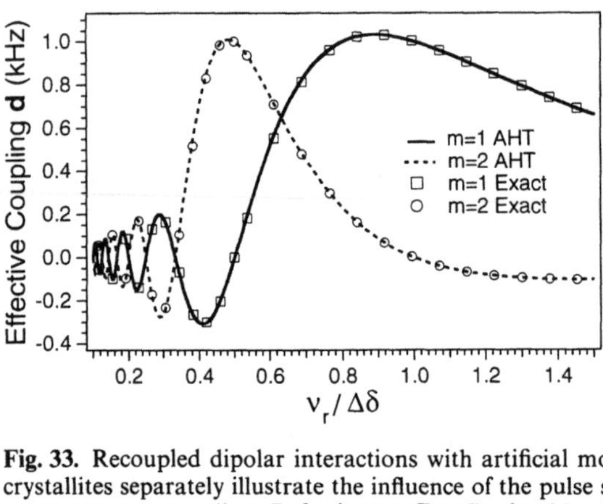

Fig. 33. Recoupled dipolar interactions with artificial model crystallites. Roughly speaking, these crystallites separately illustrate the influence of the pulse sequence on each term oscillating at $m\omega_R$ in the dipolar coupling. Referring to Eq. (5), the $M = 1$ crystallite has $\omega^D G_1 = 1$ kHz, and the $m = 2$ crystallite $\omega^D G_2 = 1$ kHz, while the other Fourier components and the angle ϕ are zero in both cases. The results of AHT in a double toggling frame are compared with exact calculations as a function of the ratio of the spinning frequency to the chemical shift difference. Reprinted from Ref. [48] with permission

in SEDRA experiments for each crystallite:

$$S(Mt_C) = \cos(\bar{\omega}^D_{eff} \cdot Mt_C) \tag{145}$$

The powder SEDRA decay can be obtained by integration of $S(Mt_C)$ over the dipolar polar angles (θ, ϕ) of the internuclear vector. In practice, experiments are conducted using the XY-8 phase cycling scheme [103] in order to reduce the influence of pulse imperfections, and therefore the signal is acquired following every eight rotor cycles. Figure 34 illustrates a dephasing trajectory and compares the experimental results with simulations [47]. To exploit the SEDRA approach for distance measurement, it is necessary to detect the transverse decay accurately. In addition, numerical simulations are necessary to calculate SEDRA decay trajectories, taking into account all relevant spin parameters and the finite duration of the RF pulses, which can be a significant effect [47]. The influence of the CSA contribution in Eq. (138) on the SEDRA frequencies can also be evaluated within the framework discussed here.

As in the case of magnetization exchange experiments at rotational resonance, spin echo recoupling experiments can be performed after an inversion of one of the spins [144] with the detection of the difference polarization:

$$\langle I_{z1} - I_{z2} \rangle (Mt_C) = \cos(2\bar{\omega}^D_{eff} \cdot Mt_C) \tag{146}$$

The effective flip-flop interaction causes exchange between the spins. Under its influence, the oscillation frequency of the polarization difference $[I_{z1} - I_{z2}]$ is twice as large as that of the coherence signal $[I_{x1} + I_{x2}]$ [59]. This distinction implies that the longitudinal approach is advantageous for the observation of weak dipolar couplings. At the same time, however, the proper treatment of longitudinal exchange experiments requires the analysis of relaxation effects, particularly the decay of zero-quantum coherence formed during the trajectory, which can slow the rate of magnetization exchange. In transverse dephasing experiments, relaxation effects are nearly isotropic, in the sense that all spin coherences decay at similar rates, so the role of relaxation is relatively unimportant.

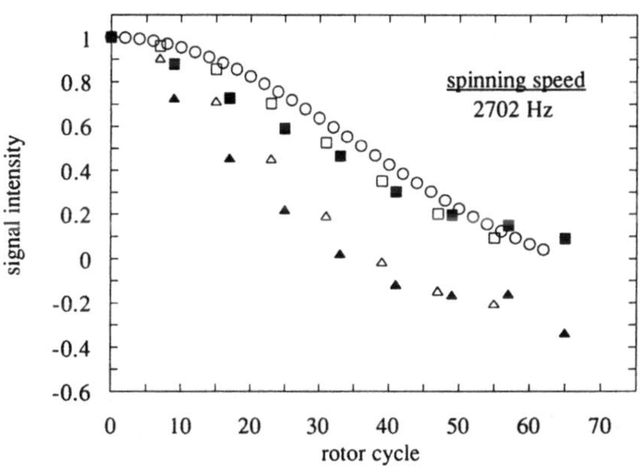

Fig. 34. SEDRA dephasing experiment employing XY-8. Results are given for (1, 3-^{13}C, ^{15}N)-D,L-alanine, whose chemical shift difference is 7.9 kHz in this experiment. The *open circles*, *squares*, and *triangles* indicate calculated trajectories with δ-function pulses and RF fields of 54 kHz and 20 kHz, respectively. The experimental results at these pulse strengths are indicated by *solid squares* and *triangles*. Reprinted from Ref. [47] with permission

The value of $\bar{\omega}^{D}_{eff}$ is not sensitive to small changes in ω_R, and is thus less sensitive to small experimental instabilities in the spinning speed than the dipolar interaction recovered in rotational resonance experiments. The spin echo approach is also less sensitive to the small departures from the rotational resonance condition that occur within an inhomogeneous lineshape. On the other hand, because of the strong dependence of $\bar{\omega}^{D}_{eff}$ on crystallite orientation, the magnitude of the recoupled flip-flop term is usually smaller than that of the dipolar coupling recovered at rotational resonance. In particular, in the limit of weak dipolar coupling, the AHT approach [48] provides the following result for the dipolar coupling recovered by a spin echo sequence at the rotational resonance condition $\Delta\omega = N\omega_R$:

$$\bar{H}_{eff} = |\omega^{D}_{N}(\theta)| \cos(N\phi)[I_{+1}I_{-2} + I_{-1}I_{+2}], \tag{147}$$

again neglecting the effect of the CSA terms. In the absence of pulses, the analogous interaction recovered at exact rotational resonance is given by:

$$\bar{H}_{eff} = |\omega^{D}_{N}(\theta)|[I_{+1}I_{-2} + I_{-1}I_{+2}] \tag{148}$$

Because of its dependence on the phase angle ϕ, the coupling recovered by spin echo sequences is attenuated relative to the interaction obtained at rotational resonance without pulses. Consequently, the rate of dipolar evolution averaged over a powder sample is attenuated by the application of π pulses. This effect can be observed in the experiments of Sodickson et al. [144] and has been noted by Gullion and Vega [47].

Another application of the recoupling effect generated by spin echoes is the RF-Driven Dipolar Recoupling (RFDR) technique, which provides an approach to the homonuclear correlation spectroscopy of dilute spins in isotopically enriched solids undergoing MAS [48, 145]. The effective dipolar flip-flop term

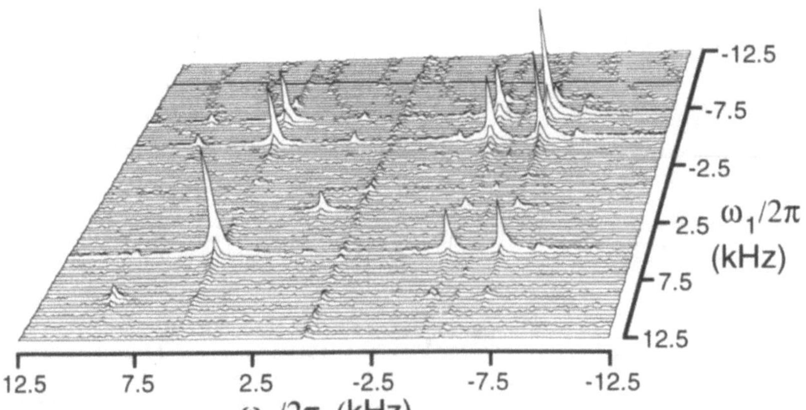

Fig. 35. 2D RFDR Spectrum. The homonuclear correlation spectrum of triply-^{13}C-labeled D,L-alanine is shown with a 8.6 ms mixing time at 3.72 kHz spinning speed. Reprinted from Ref. [48] with permission

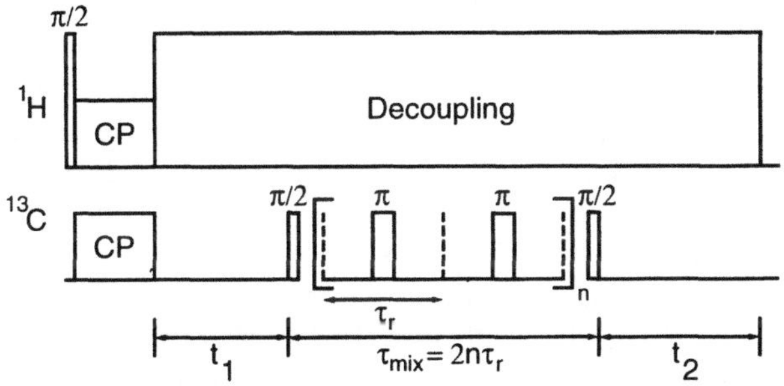

Fig. 36. 2D RFDR pulse sequence. The pulse sequence used to obtain the spectrum in Fig. 35 is illustrated. The $\pi/2$ pulses store the magnetization along the magnetic field for longitudinal exchange with π pulses. Reprinted from Ref. [48] with permission

generated by the spin echo sequences is applicable to two-dimensional longitudinal exchange experiments, since observable spin polarizations are directly coupled by the zero-quantum coherence [48, 145]. The basic SEDRA experiment does not couple observable magnetizations of the two spins together. A 2D exchange experiment of triply-^{13}C-labeled alanine, acquired using the RFDR technique, is shown in Fig. 35 [48]. The pulse sequence is illustrated in Fig. 36. Since the experiment reintroduces the dipolar couplings among all three spins simultaneously, cross peaks are readily observed connecting the three neighboring homonuclear spins. The longitudinal exchange method is an efficient approach for stimulating and observing dipolar recoupling over a broad range of chemical shifts, and it is therefore useful for the application of homonuclear recoupling techniques to systems involving more than two spins.

5 Applications

In order to emphasize the practical utility of the techniques discussed in this article, we provide a brief discussion of two examples in which dipolar recoupling methods have been used to investigate the structure of proteins. The first example is the application of rotational resonance to the measurement of homonuclear interatomic distances [134, 136, 137] in the retinal binding pocket of the membrane protein bacteriorhodopsin (bR) [146]. The rotational resonance technique has been successfully applied to the study of both the ground state [134, 136] and a photo-intermediate, the M_{412} state of the bR photocycle [137]. In the second example, using REDOR experiments, Christensen and Schaefer [100] have measured heteronuclear interatomic distances in a lyophilized ternary enzyme-substrate complex that had resisted investigation by X-ray crystallography and solution NMR spectroscopy. REDOR experiments have also been applied to the study of several other biological molecules [107–109, 112].

Bacteriorhodopsin, which has a molecular weight of 26.6 kDa, is suitable neither for study by solution NMR spectroscopy, since it is insoluble, nor by X-ray crystallography, since it does not form three-dimensional crystals. A structure based on electron microscopy has been proposed by Henderson [147], which serves as a point of departure for NMR investigations [62, 148, 149]. Rotational resonance has been used to measure a number of internuclear distances in biological molecules [133–138]. Employing magnetization exchange experiments at rotational resonance, Thompson et al. [134] have investigated the conformations of the Schiff base linkage between the retinal and the side chain of Lys_{216} in the dark-adapted state of bR. In the dark-adapted form, two conformers, bR_{555} and bR_{568} (denoted by their optical absorption maxima), whose structures near the Schiff base linkage are illustrated in Fig. 37, exist in an approximately 60:40 ratio at thermal equilibrium. In the M_{412} state of the bR photocycle, in which the Schiff base nitrogen is deprotonated, two conformations are possible at the Schiff base linkage: 13-*cis*, 15-*anti* and 13-*cis*, 14-*syn*.

MAS spectra of a sample of bR, ^{13}C-labeled at the 14-*C* position of the retinal and the ε-*C* of the Lys_{216} sidechain, are shown in Fig. 38 [149], demonstrating the clear resolution of the spin labels found in the dark-adapted, light-adapted, and M intermediate states, respectively. In each component of the dark-adapted state, the distance between the nuclei was measured by the observation of exchange at the $N = 1$ rotational resonance condition. Since rotational resonance is a frequency-selective method, a separate experiment was conducted for each species. Figure 39 illustrates the magnetization exchange trajectories. The internuclear distances that were determined provide convincing evidence that the Schiff base conformation in bR_{555} is 13-*cis*, 15-*syn* (3.0 ± 0.2 Å) and the configuration in bR_{568}, 13-*trans*, 15-*anti* (4.1 ± 0.3 Å). Moreover, the measurement in the bR_{568} component provides a guide to the molecular size which can be approached with the technique. Specifically, since the protein of MW = 26.6 kDa resides in a membrane which is 25% lipid, bR itself has an effective MW ≅ 35 kDa. Furthermore, the bR_{568} component corresponds to 0.4 of a single ^{13}C, and therefore the distance measurement in this component corresponds to studying a protein of effective MW ≅ 85 kDa. More recently, Lakshmi et al. [137] have conducted similar experiments to measure the molecular configuration in the cryogenically trapped M intermediate state of the photocycle. In this case, it was determined that the retinal and C = N linkage adopts the 13-*cis*, 15-*anti* configuration. These experiments on the M_{412} state

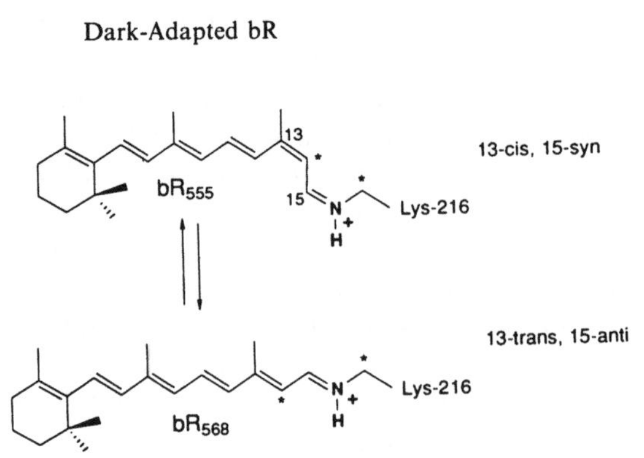

Fig. 37. Conformations and Schiff base linkage between the protein and the retinal in dark-adapted bR. The two components exist in an approximately 60:40 ratio. The ^{13}C-labeled nuclei near the Schiff base linkage are indicated with asterisks. Reprinted from Ref. [134] with permission

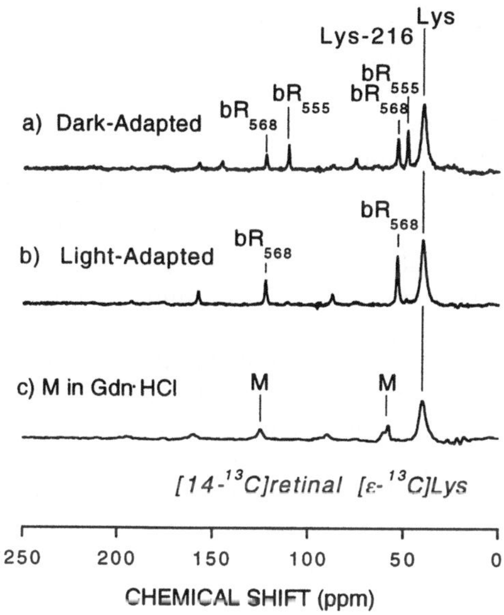

Fig. 38. MAS Spectra of bR, ^{13}C-labeled at the ε-Lys and 14-Ret positions. The NMR spectra of the labeled nuclei are selectively observed by difference experiments, in which the spectrum of a natural abundance sample is acquired and subtracted from the spectrum of the labeled sample. Reprinted from Ref. [149] with permission

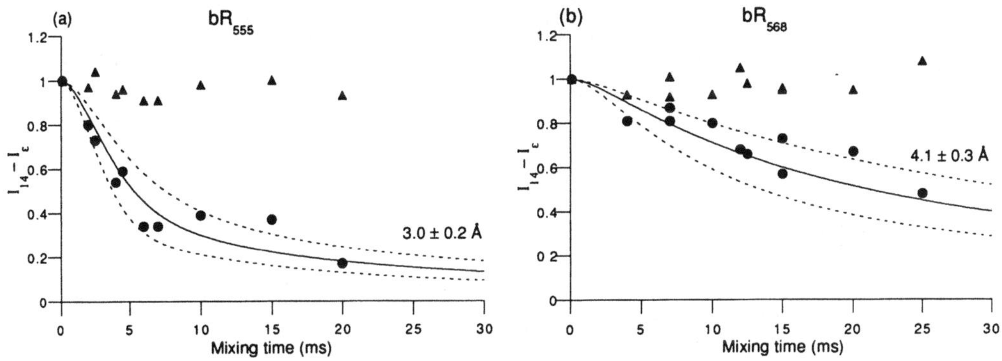

Fig. 39a, b. Rotational resonance exchange in bR. Magnetization exchange is plotted at exact rotational resonance (*solid circles*) and somewhat away from rotational resonance (*solid triangles*). The *solid lines* are the results of numerical simulations with the indicated distances, and the *dashed lines* indicate simulated results with distances that deviate from the values corresponding to the optimal match by the indicated ranges of error. The experiments required the spinning speeds: **a** 6.250 kHz and **b** 6.950 kHz, in order to match the $N = 1$ rotational resonance condition. Reprinted from Ref. [134] with permission

demonstrate the potential of rotational resonance and other solid state recoupling methods for the accurate measurement of interatomic distances in reaction intermediates.

Christensen et al. [100] employed the REDOR technique to measure ^{31}P-^{13}C distances in the ternary complex of shikimate 3-phosphate (S3P) and *N*-(phosphonomethyl)-(1-^{13}C)-glycine (glyphosate) bound to the enzyme 5-

enolpyruvylshikimate-3-phosphate (EPSP). Because of the relatively large gyromagnetic ratios of ^{31}P and ^{13}C, heteronuclear spin pairs involving these nuclei have reasonably large dipolar couplings. Consequently, it was possible to measure two exceptionally long ^{31}P-^{13}C distances within the ternary enzyme-substrate complex. In particular, it was determined that the distance between the labeled ^{13}C and the intramolecular ^{31}P in glyphosate is 5.6 ± 0.2 Å, suggesting that the glyphosate adopts an extended conformation in the complex. In addition, the intermolecular distance of 7.2 ± 0.4Å between the ^{13}C in glyphosate and the shikimate 3-phosphate indicates the proximity of the two components in the ternary complex.

The ^{31}P MAS spectra of the ^{13}C-labeled complex and its natural abundance analogue are shown in Fig. 40 and compared with the REDOR difference spectra. In contrast to rotational resonance, the REDOR experiment recouples interactions without dependence on frequency, so the dephasing trajectories of both ^{31}P resonances were observed in a single experiment. Similar dephasing experiments were performed on a natural abundance sample of the complex in order to measure and subtract the contribution to the dipolar trajectories from the background of naturally occurring ^{13}C spins. Figure 41 illustrates a possible structure of the S3P-glyphosate complex bound to EPSP that is consistent with the two constraints of high accuracy obtained from REDOR experiments.

A particularly important point illustrated by these applications is that MAS experiments are applicable to molecular states that are inaccessible by other approaches. In addition to providing new capabilities in examining polycrystalline and amorphous states, dipolar recoupling experiments often provide

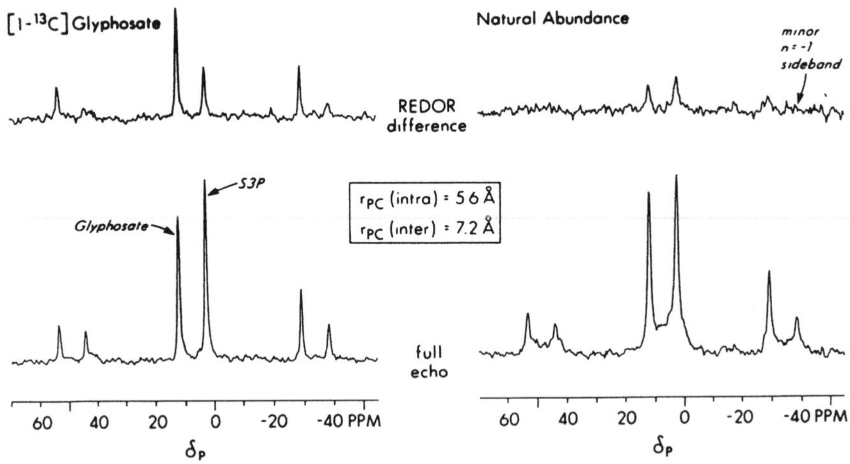

Fig. 40. ^{31}P REDOR spectra of the ternary complex of S3P and glyphosate bound to EPSP synthase. The complete specta and the REDOR differences are shown after 64 rotor periods of dephasing at 5 kHz spinning speed. The dipolar interactions and observed internuclear distances involve coupling between the ^{31}P nuclei and the ^{13}C label in glyphosate. Reprinted from Ref. [100] with permission

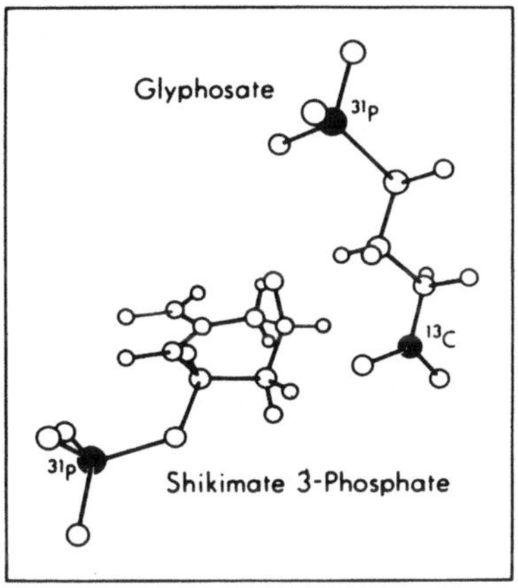

Fig. 41. Possible structure of the ternary complex of glyphosate and S3P in EPSP synthase. The *solid spheres* indicated the atoms involved in the REDOR experiment. Reprinted from Ref. [100] with permission

interatomic distances that are much more accurate than those obtained from Nuclear Overhauser Effect Spectroscopy (NOESY) in the solution state [150–152]. On the other hand, to date, most experiments have been limited to the case of two labeled spins. With this approach, a limited number of interatomic distances can be determined to relatively high accuracy. However, the two-dimensional spectrum shown in Fig. 35 illustrates the potential of dipolar recoupling methods to determine many distances simultaneously. It is this approach or a similar one which will probably find wide applicability.

6 Conclusions

In this review, we have examined a variety of useful techniques that achieve the recoupling of through-space dipolar interactions between dilute spins into MAS experiments. In the case of synchronous detection, effective Hamiltonians under MAS and multiple pulse sequences can be derived by the application of either AHT or Floquet Theory to the time-dependent Hamiltonian in the toggling frame. The Floquet approach allows the evaluation of spin evolution at all times. It is therefore sometimes preferable to use Floquet Theory over the AHT approach. For example, in the case of rotational resonance, it is useful to determine the FID at all times in order to obtain a complete picture of the MAS spectrum. Otherwise, the choice between the two approaches is a matter of taste or convenience. Recoupled dipolar interactions can be used in a variety of ways in MAS experiments and have found applicability in two-dimensional correlation spectroscopy [48, 116, 145], filtering experiments [46, 114, 115, 132], and most important, in the measurement of internuclear distances in selected dilute spin pairs.

We conclude by noting that the field of dipolar recoupling experiments continues to evolve rapidly. New methods that are based upon the essential ideas reviewed in this article are still under development in various laboratories, and there is an expanding interest in applying recoupling techniques to structural problems in polycrystalline and amorphous solids. MAS experiments that reveal dipolar couplings provide a new window into the structure of condensed phases at the molecular level. These methods are particularly significant for problems that are difficult to investigate by other techniques.

Acknowledgments. The authors wish to thank J. Schaefer, T. Gullion, R. Tycko, S. Opella, M. H. Levitt, A. E. McDermott, D. P. Raleigh, F. Creuzet, A. Kolbert, L. K. Thompson, J. H. Ok, O. Weintraub, A. Schmidt, B. Sun, and J. M. Griffiths for helpful discussions. In addition, the assistance of J. Schaefer, T. Gullion, R. Tycko, S. Opella, and J. M. Griffiths with the preparation of the manuscript is greatly appreciated. A. E. B. was the recipient of an NSF Predoctoral Fellowship. Research in magic angle spinning NMR spectroscopy at the Francis Bitter National Magnet Laboratory has been supported by NIH grants RR-00995, GM-23403, GM-25505, and GM-23289.

7 References

1. Pake GE (1948) J Chem Phys 16: 327
2. Hester RK, Ackerman JL, Neff BL, Waugh JS (1976) Phys Rev L 36: 1081
3. Stoll ME, Vega AJ, Vaughan RW (1976) J Chem Phys 65: 4093
4. Linder M, Höhener A, Ernst RR (1980) J Chem Phys 73: 4959
5. Munowitz MG, Huang TH, Dobson CM, Griffin RG (1984) J Magn Res 57: 56
6. Herzog B, Hahn EL (1956) Phys Rev 103: 148
7. Kaplan DE, Hahn EL (1958) J Phys Radium 19: 821
8. Kaplan DE, Hahn EL (1957) Bull Am Phys Soc 2: 384
9. Emshwiller M, Hahn EL, Kaplan DE (1960) Phys Rev 118: 414
10. Slichter CP (1990) Principles of magnetic resonance. Springer, Berlin Heidelberg New York
11. Shore SE, Amsermet J-P, Slichter CP, Sinfelt JH (1987) Phys Rev L 58: 953
12. Yannoni CS, Kenderick RD (1981) J Chem Phys 74: 747
13. Engelsberg M, Yannoni CS (1990) J Magn Res 88: 393
14. Lizak MJ, Gullion T, Conradi MS (1991) J Magn Res 91: 254
15. Gullion T, Schaefer J (1989) J Magn Res 81: 196
16. Gullion T, Schaefer J (1989) Adv Mag Res 13: 58
17. Andrew ER, Bradbury A, Eades RG (1958) Nature (London) 182: 1659
18. Lowe IJ (1959) Phys Rev L 2: 285
19. Schaefer J, Stejskal EO (1976) J Am Chem S 98: 1030
20. Lippmaa E, Alla M, Tuherm T (1976) In: Brunner, H Hausser, RH, Schweitzer, D (eds) Proc. XIX Congress Ampere. Heidelberg, Group Ampere, p 113
21. Andrew ER, Hinshaw WS, Tiffen RS (1974) J Magn Res 15: 191
22. Maricq M, Waugh JS (1979) J Chem Phys 70: 3300
23. Pines A, Gibby MG, Waugh JS (1973) J Chem Phys 59: 569
24. Munowitz MG, Griffin RG (1982) J Chem Phys 76: 2848
25. Munowitz MG, Griffin RG (1983) J Chem Phys 78: 613
26. Schaefer J, McKay RA, Stejskal EO, Dixon WT (1983) J Magn Res 52: 123
27. Kolbert AC, Raleigh DP, Levitt MH, Griffin RG (1989) J Chem Phys 90: 679

28. There is, however, at least one exception. Maricq and Waugh discussed the case of a linear array of homonuclear spins undergoing MAS [22]. In this system, rotational echoes are formed because the complete dipolar Hamiltonian is self-commuting under the modulation.
29. Haeberlen U, Waugh JS (1968) Phys Rev 175: 453
30. Haeberlen U (1976) High resolution NMR in solids: Selective averaging. Academic, New York
31. Shirley JM (1965) Phys Rev B 138: 979
32. Dion DR, Hirschfelder JO (1976) Adv Chem Phys 35: 265
33. Zur Y, Levitt MH, Vega S (1983) J Chem Phys 78: 5293
34. Zax DB, Goelman G, Abramovich D, Vega S (1990) Adv Magn Res 14: 219
35. Oas TG, Griffin RG, Levitt MH (1988) J Chem Phys 89: 692
36. Levitt MH, Oas TG, Griffin RG (1988) Isr J Chem 28: 271
37. Schmidt A, Vega S (1992) Isr J Chem 32: 215
38. Diverdi JA, Opella SJ (1982) J Am Chem S 104: 1761
39. Pan Y, Gullion T, Schaefer J (1990) J Magn Res 90: 330
40. Raleigh DP, Levitt MH, Griffin RG (1988) Chem Phys L 146: 71
41. Levitt MH, Raleigh DP, Creuzet F, Griffin RG (1990) J Chem Phys 92: 6347
42. Columbo MG, Meier BH, Ernst RR (1988) Chem P Lett 146: 189
43. Kubo A, McDowell CA (1988) J Chem Soc Faraday Trans 84: 3713
44. Maas WEJR, Veeman WS (1988) J Chem Phys L 149: 170
45. Tycko R, Dabbagh G (1990) Chem Phys L 173: 461
46. Tycko R, Smith SO (1992) J Chem Phys 98: 932
47. Gullion T, Vega S (1992) Chem Phys L 194: 423
48. Bennett AE, Ok JH, Griffin RG, Vega S (1992) J Chem Phys 96: 8624
49. Mehring M (1976) Principles of High Resolution NMR in Solids. Springer-Verlag, Berlin, Germany
50. Mehring M, Pines A, Rhim W-K, Waugh JS (1971) J Chem Phys 54: 3239
51. Olejniczak ET, Vega S, Griffin RG (1984) J Chem Phys 81: 4804
52. Wokaun A, Ernst RR (1977) J Chem Phys 67: 1752
53. Vega S (1978) J Chem Phys 68: 5518
54. Sakurai JJ, Tuan SF (1985) Modern quantum mechanics. Benjamin/Cummings, Menlo Park, California
55. Stejskal EO, Schaefer J, McKay RA (1977) J Magn Res 25: 569
56. Maricq M, Waugh JS (1977) Chem Phys L 47: 327
57. Waugh JS, Maricq MM, Cantor R (1978) J Magn Res 29: 183
58. Herzfeld J, Berger AE (1980) J Chem Phys 73: 6021
59. Schmidt A, Vega S (1992) J Chem Phys 96: 2655
60. Munowitz MG (1982) Ph.D. Thesis, Harvard University, Cambridge, Massachusetts
61. Levitt MH (1989) J Magn Res 82: 427
62. Smith SO, Griffin RG (1988) Ann Rev Phys Chem 39: 511
63. Herzfeld J, Roufosse A, Haberkorn RA, Griffin RG, Glimcher MJ (1980) Philos Trans R Soc (London) B 289: 459
64. Vega S, Olejniczak ET, Griffin RG (1984) J Chem Phys 80: 4832
65. Kubo A, McDowell CA (1990) J Chem Phys 92: 7156
66. Nakai T, McDowell CA (1992) J Chem Phys 96: 3452
67. Weintraub O, Vega S (1994) J Magn Res 105: 245
68. Maricq MM (1990) Adv Magn Res 14: 151
69. Maricq MM (1982) Phys Rev B 25: 6622
70. Llor A (1992) Chem Phys L 199: 383
71. Caravatti P, Bodenhausen G, Ernst RR (1983) J Magn Res 55: 88
72. Waugh JS (1982) J Magn Res 50: 30
73. Magnus W (1954) Comm Pure Math 7: 649
74. Pechukas P, Light JC (1966) J Chem Phys 44: 3897
75. Evans WAB (1968) Ann Phys 48: 72
76. Shirley MJ (1963) Ph.D. Thesis, California Institute of Technology, Pasadena, California
77. Munowitz MG, Griffin RG, Bodenhausen G, Huang T (1981) J Am Chem S 103: 2529
78. Miura H, Terao T, Saika A (1986) J Chem Phys 85: 2458
79. Raleigh DP, Olejniczak ET, Griffin RG (1988) J Chem Phys 89: 1333
80. Kolbert AC, Levitt MH, Griffin RG (1989) J Magn Res 85: 42
81. Kolbert AC, Raleigh DP, Griffin RG (1989) J Magn Res 82: 483
82. Munowitz MG, Griffin RG, Bodenhausen G, Huang TH (1982) J Am Chem S 103: 2529

83. Roberts JE, Harbison GS, Munowitz MG, Herzfeld J, Griffin RG, (1987) J Am Chem S 109:4163
84. Anderson PW, Freeman R (1962) J Chem Phys 37:85
85. Munowitz WG, Aue WP, Griffin RG (1982) J Chem Phys 77:1686
86. Schaefer J, Stejskal EO, McKay RA, Dixon WT (1984) Macromolec 17:1479
87. Webb GG, Zilm KW (1989) J Am Chem S 111:2455
88. Waugh JS, Huber LM, Haeberlen U (1968) Phys Rev L 20:180
89. Mansfield P (1991) J Phys C 4:1444
90. Rhim WK, Elleman DD, Vaughan RW (1973) J Chem Phys 59:3740
91. Lee M, Goldberg WI, (1965) Phys Rev A 140:1261
92. Burum DP, Rhim W-K (1979) J Chem Phys 71:944
93. Burum DP, Linder M, Ernst RR (1981) J Magn Res 44:173
94. Gullion T, Policks MD, Schaefer J (1988) J Magn Res 80:553
95. Bork V, Gullion T, Hing A, Schaefer J (1989) J Magn Res 88:523
96. Schaefer J, Stejskal EO, McKay RA, Dixon WT (1984) J Magn Res 57:85
97. Schaefer J, Sefcik MD, Stejskal EO, McKay RA, Dixon WT, Cais RE (1984) Macromolec 17:1107
98. Schaefer J, Stejskal EO, Perchak DS, Skolnick J, Yaris R (1985) Macromolec 18:368
99. Garbow JR, Jacob GS, Stejskal EO, Schaefer J (1989) Biochem 28:1362
100. Christensen AM, Schaefer J (1993) Biochem 32:2868
101. Gullion T, Conradi MS (1990) J Magn Res 86:39
102. Pan Y, Schaefer J (1990) J Magn Res 90:341
103. Gullion T, Baker DB, Conradi MS (1990) J Magn Res 89:479
104. Gullion T, Schaefer J (1991) J Magn Res 92:439
105. Garbow JR, Gullion T (1991) J Magn Res 95:442
106. Garbow JR, Gullion T (1992) Chem Phys L 192:71
107. Marshall GR, Beusen DD, Kociolek K, Redlinski AS, Leplawy MT, Pan Y, Schaefer J (1990) J Am Chem S 112:963
108. Christensen AM, Schaefer J, Kramer KJ, Morgan TD, Hopkins TL (1991) J Am Chem S 113:6799
109. Garbow JR, McWherter CA (1993) J Am Chem S 115:238
110. Holl SM, Marshall GR, Beusen DD, Kociolek K, Redlinski AS, Leplawy MT, McKay RA, Vega S, Schaefer J (1992) J Am Chem S 114:4830
111. Schmidt A, McKay RA, Schaefer J (1992) J Magn Res 96:644
112. Holl SM, McKay RA, Gullion T, Schaefer J (1990 J Magan Reson 89:620
113. Morris J Freeman R (1979) J Am Chem S 101:760
114. Hing A, Vega S, Schaefer J (1992) J Magn Res 96:205
115. Hing A, Vega S, Schaefer J (1993) J Magn Res A 103:151
116. Fyfe CA, Meuller KT, Gondey H, Wong-Moon KC (1992) Chem Phys L 199:198
117. Andrew ER, Clough S, Farnell LF, Gledhill TA, Roberts I (1966) Phys Lett 21:505
118. Raleigh DP, Harbison SG, Neiss TG, Roberts JE, Griffin RG (1987) Chem Phys L 138:285
119. Meier BH, Earl WL (1987) J Am Chem S 109:7937
120. Portis AM (1953) Phys Rev 91:1071
121. Andrew ER, Bradbury A, Eades RG, Wynn VT (1963) Phys Lett 4:99
122. Gan ZH, Grant DM (1989) Mol Phys 67:1419
123. Challoner R, Nakai T, McDowell CA (1991) J Chem Phys 94:7038
124. Challoner R, Nakai T, McDowell CA (1991) Chem Phys L 180:13
125. Challoner R, Nakai T, McDowell CA (1991) J Magn Res 94:433
126. Nakai T, Challoner R, McDowell CA (1991) Chem Phys L 180:13
127. Nakai T, McDowell CA (1992) Mol Phys 77:569
128. Kubo A, Root A, McDowell CA (1990) J Chem Phys 93:5462
129. Bodenhausen G, Freeman R, Morris GA (1976) J Magn Res 23:171
130. Raleigh DP, Creuzet F, Das Gupta SK, Levitt MH, Griffin RG (1989) J Am Chem S 111:4502
131. Nielsen NC, Creuzet F, Griffin RG (1993) J Magn Res A 103:245
132. Nielsen NC, Creuzet F, Griffin RG, Levitt MH (1992) J Chem Phys 96:5668
133. Spencer RGS, Halverson KJ, Auger M, McDermott AE, Griffin RG, Lansbury PT (1991) Biochem 30:10382
134. Thompson LK, McDermott AE, Raap J, van der Wielen CM, Lugtenburg J, Herzfeld J, Griffin RG (1992) Biochem 31:7931

135. Peerson OB, Yoshimura S, Hojo H, Aimoto S, Smith SO (1992) J Am Chem S 114:4332
136. Creuzet F, McDermott AE, Gebhard R, van der Hoef K, Spyker-Assink MB, Herzfeld J, Lugtenberg J, Levitt MH, Griffin RG (1991) Science 257:783
137. Lakshmi KV, Auger M, Raap J, Lugtenburg J, Griffin RG, Herzfeld J (1993) J Am Chem S 115:8515
138. McDermott AE, Creuzet F, Griffin RG, Zawadzke L, Ye QZ, Walsh CT (1990) Biochem 29:5767
139. Tycko R (1988) Phys Rev L 60:2734
140. Zax DB, Bielecki A, Zilm KW, Pines A, Weitekamp DP (1985) J Chem Phys 83:4877
141. Fujiwara T, Ramamoorthy A, Nagayama K, Hoika K, Fujito T (1993) Chem Phys L 212:81
142. Ramamoorthy A, Fujiwara T, Nagayama K (1993) J Magn Res A 104:366
143. Rhim W-K, Pines A, Waugh JS (1971) Phys Rev B 3:684
144. Sodickson DK, Levitt MH, Vega S, Griffin RG (1993) J Chem Phys 98:6742
145. Ok JH, Spencer RGS, Bennett AE, Griffin RG (1992) Chem Phys L 197:389
146. Khorana HG (1988) J Bio Chem 263:7439
147. Henderson R, Baldwin JM, Ceska TA, Zemlin F, Beckmann E, Downing KH (1990) J Mol Bio 213:899
148. Harbison GS, Smith SO, Pardoen JA, Mulder PPJ, Lugtenburg J, Herzfeld J, Mathies RA, Griffin RG (1984) Proc Natl Acad Sci USA 81:1706
149. Farrar MR, Lakshmi KV, Smith SO, Brown RS, Raap J, Lugtenburg J, Griffin RG, Herzfeld J (1993) Biophys J 65:310
150. Jeener J, Meier BH, Bachmann P, Ernst RR (1979) J Chem Phys 71:4546
151. Ernst RR, Bodenhausen G, Wokaun A (1987) Principles of nuclear magnetic resonance in one and two dimensions. Clarendon, Oxford, England
152. Wüthrich K (1986) NMR of proteins and nucleic acids. John Wiley, New York

Solid-State NMR Line Narrowing Methods for Quadrupolar Nuclei: Double Rotation and Dynamic-Angle Spinning

B. F. Chmelka[1] and J. W. Zwanziger[2]

[1] Department of Chemical and Nuclear Engineering, University of California, Santa Barbara, California 93106-5080, USA

[2] Department of Chemistry, Indiana University, Bloomington, IN 47405, USA

Table of Contents

Recent theoretical advances coupled with innovative engineering have led to rapid progress in the development of new methods for obtaining high-resolution NMR spectra of quadrupolar nuclei in solids. This paper provides a critical review of the theory, implementation, and applications of several new multiple-angle sample reorientation techniques. The purpose of these methods is to remove anisotropic broadening due to second-order effects of the local interactions, in particular of the quadrupole interaction. After establishing a foundation for first-order averaging schemes (e.g., Magic-Angle Spinning) based on Static Perturbation Theory, second-order expressions are derived. This treatment shows the origin and symmetry of the various time-dependent, multiple-angle sample reorientation trajectories that have been developed to remove second-order anisotropic broadening. Double Rotation and Dynamic-Angle Spinning experiments are emphasized, with accompanying discussions of their feasibilities and limitations toward providing high-resolution solid-state NMR spectra of quadrupolar nuclei.

NMR Basic Principles and Progress, Vol. 33

1 Introduction

Among the features of nuclear magnetic resonance (NMR) spectroscopy that have made it such a versatile and powerful collection of analytical methods is the freedom to manipulate independently both spin and spatial components of the nuclear spin interaction tensors. This imparts enormous flexibility and selectivity to the type of structural and dynamical information that can be obtained about a chemical system at a molecular level. Depending on the information desired, experiments can be designed to measure, remove, or correlate one or more of the principal spin interactions, including the chemical shift, spin–spin scalar (J) coupling, electric quadrupole coupling, and magnetic dipole–dipole coupling, each of which depends on molecular orientation. Often one of the chief aims of such intervention is to enhance spectral resolution, and thereby augment the information content of the NMR spectrum, through judicious control of time-dependent radiofrequency ($\mathbf{B}_1$) fields and/or manipulation of sample orientation with respect to a large static field ($\mathbf{B}_0$). In low-viscosity liquids, fast isotropic molecular motion averages all anisotropic interactions, so that narrow line widths are readily obtained for each distinguishable site in a molecular system. Narrow spectral lines in these systems reflect, for example, the isotropic value of each site-specific chemical shift tensor (modified potentially by an isotropic J-coupling contribution as well), which is averaged over all possible orientations by rapid molecular motion on the NMR time scale. Such an isotropic sampling criterion is equivalent to averaging all nuclear spin interactions over a sphere, whereby only those contributions that are independent of molecular orientation remain, namely the isotropic terms.

Whereas fast isotropic molecular motion eliminates anisotropic spectral broadening in liquids, the absence of rapid molecular tumbling in most solids results in broadening influences from anisotropic interactions, which can significantly limit spectral resolution. Manipulation of the spin component of the nuclear spin interaction has led to the development of powerful multiple-pulse solid-state NMR techniques that have produced tremendous gains in sensitivity and resolution, among the most omnipresent and successful examples being cross-polarization and decoupling [1–4]. Alternatively, control of the spatial component provides separate opportunities for the experimenter to enhance spectral resolution, by averaging orientation-dependent spatial terms to eliminate anisotropic contributions to the spectrum. Fortunately, such broadening influences can be removed under less stringent conditions than the isotropic averaging process observed in simple liquids, allowing one to direct the sample through a much smaller subset of well-chosen orientations to eliminate the anisotropic terms. All NMR sample reorientation techniques rely on the well-defined directional dependencies of anisotropic nuclear spin interactions, which transform, as we shall see, as zeroth- and second-rank tensors that can be averaged efficiently over a manageable subset of molecular orientations. This review aims to compare and contrast the ways in which several new sample reorientation NMR techniques confront this issue, and in so doing examine the information provided by the different methods, along with the relative utilities and merits of each approach. We will focus here on the theory and experimental implementation of several newly developed methods for improving spectral resolution of quadrupolar nuclei in solids through elimination of multiple-rank broadening. It is intended that this will complement and extend other discussions that have recently appeared on this topic [5–8]. Emphasis will be placed on new Double Rotation and Dynamic-Angle Spinning techniques, which are proliferating widely and which provide exceptional new opportunities for analyzing local structural and dynamical features of complicated inorganic solids.

2 Theory of Rapid Sample Reorientation

2.1 *Averaging First-Order Interactions*

As shown over 30 years ago by Andrew et al. [9] and Lowe [10], anisotropic spectral broadening from weak dipole–dipole couplings can be averaged away by spinning a sample rapidly about an angle inclined at 54.74° (the so-called "magic-angle") with respect to the static $\mathbf{B}_0$ magnetic field. This is an example of the fact that it is not necessary to sample all possible orientations to average a particular anisotropic interaction to zero; rather, for a given interaction, a certain well-prescribed subset of orientations is sufficient. The mathematical foundation for such averaging procedures is well-established and based on the general transformation properties of second-rank tensors under rotations [2, 3]. To first order, the chemical shift, scalar (*J*) coupling, and quadrupolar coupling interactions transform in the same way as dipole–dipole couplings, and so Magic-Angle Spinning (MAS) is an effective averaging strategy for all of them.

In systems with strong quadrupolar interactions, however, higher-order effects become important, with the result that Magic-Angle Spinning fails to remove completely anisotropic spectral broadening. To appreciate the ingenious experimental strategies invented to average higher-order quadrupolar effects, it is helpful to understand their theoretical foundation. As this follows in a direct way from the first-order treatment, we summarize and use examples of first-order averaging as the logical basis from which higher-order averaging methods may be developed and best understood.

In general, an ensemble of spins subject to an internal interaction λ and a static external magnetic field can be described by the Hamiltonian (see Ref. [2])

$$\mathscr{H} = \mathscr{H}_Z + \mathscr{H}_\lambda \tag{1}$$

$$= -\omega_0 I_z + \omega_\lambda \sum_{l=0}^{2} \sum_{m=-l}^{l} (-1)^m R^\lambda_{l,-m} T^\lambda_{l,m}, \tag{2}$$

resulting in

$$\mathscr{H} = -\omega_0 I_z + \omega_\lambda \left[R^\lambda_{0,0} T^\lambda_{0,0} + \sum_{m=-2}^{2} (-1)^m R^\lambda_{2,-m} T^\lambda_{2,m} \right]. \tag{3}$$

In these expressions, ω_0 is the Larmor frequency of nuclear precession in the static $\mathbf{B}_0$ field, ω_λ is a constant (assumed to be $\ll \omega_0$) that describes the strength of a particular interaction λ, and $R^\lambda_{l,-m}$ and $T^\lambda_{l,m}$ are the elements of the irreducible spherical spatial and spin tensors, respectively. For the interactions considered here, primarily chemical shift (assumed symmetric) and quadrupolar, the only non-zero components are $l = 0, 2$ [2, 11].

One approach frequently used to compute the NMR signal from the Hamiltonian in Eq. (3) is Average Hamiltonian Theory (AHT). As is well-known, AHT is an approximation scheme for calculating the propagator of a system after n periods of its motion. It is therefore particularly useful for experiments that use stroboscopic sampling. Although the sample-spinning experiments to be discussed below, including MAS, are almost never performed this way, AHT has nevertheless been used to describe them. The alternative description for these experiments is Static Perturbation Theory (SPT). To first order, the AHT propagator agrees with that obtained from SPT for sample-spinning experiments [12]. However, for the experiments discussed here, which include higher-order effects and non-stroboscopic sampling, AHT, in contrast to SPT, can lead to erroneous predictions [13]. Our theoretical treatment therefore emphasizes the SPT description of averaging experiments that are designed to eliminate broadening effects from anisotropic higher-order interactions.

Goldman et al. [13] have pointed out that the SPT expansion of the typical NMR Hamiltonian (Eq. 3) can be conveniently expressed in terms of spherical tensor operators when the Zeeman interaction $-\omega_0 I_z$ is dominant. This, then, is the zeroth-order term, and the first-order term is just that part of $\mathscr{H}_\lambda$ that commutes with it. The first-order SPT result is then

$$\mathscr{H} = \mathscr{H}_Z + \mathscr{H}_\lambda \approx -\omega_0 I_z + \omega_\lambda R^\lambda_{0,0} T^\lambda_{0,0} + \omega_\lambda R^\lambda_{2,0} T^\lambda_{2,0}, \tag{4}$$

where use has been made of the commutation rule

$$[I_z, T_{l,m}] = m T_{l,m}. \tag{5}$$

The anisotropic broadening produced by the above Hamiltonian (Eq. 4) arises from the last term. One approach to averaging this anisotropy focuses on the spin degrees of freedom, expressed through $T^\lambda_{2,0}$, which can be manipulated according to standard multiple-pulse strategies. Here, we will instead focus on the irreducible spatial tensor components $R^\lambda_{l,m}$ and their behavior under finite rotations, which can be conveniently expressed in terms of Wigner rotation matrices $D(\alpha,\beta,\gamma)$ and the interaction spatial tensor components $\rho^\lambda_{l,m}$ in the principal axes system [2, 14, 15]. Together, these relate the effect of an interaction λ in the principal axes system (PAS) of the molecular system under scrutiny to the laboratory frame in which manipulations of the sample (i.e., sample rotations) will be carried out. Precisely, we have

$$R^{\lambda}_{1,m} = \sum_{m'=-1}^{1} D^{(1)}_{m',m}(\Omega)\rho^{\lambda}_{1,m'}, \tag{6}$$

where $\Omega = (\alpha, \beta, \gamma)$ are the Euler angles required to effect transformation of the PAS of the interaction λ into the laboratory frame and $D^{(l)}_{m',m}(\Omega) = \langle l, m' | D(\Omega) | l, m \rangle$ are the matrix elements of Wigner rotation operators $D(\Omega)$. For any given crystallite in the sample, Eq. (6) in conjunction with Eq. (4) can be used to compute the resonance frequency. For a polycrystalline or disordered sample, an average over all PAS orientations must be carried out, which leads to the usual powder patterns of solid-state NMR.

For the case of a sample rotating at a frequency ω_r about an axis inclined at angle β with respect to the static field, the relationship of an arbitrary PAS orientation to the laboratory frame will clearly be time-dependent. Such a complication, however, is a minor one and describable in a straightforward way by using two transformations. The first is time-independent, and relates the PAS of an interaction λ in an arbitrary crystallite to a sample-fixed frame. We will denote the Euler angles describing this transformation by Ω_{PAS}. The second transformation is time-dependent, and relates the sample-fixed frame to the laboratory frame. This transformation is described by the Euler angles $\Omega_r(t) = (0, \beta, \omega_r t + \gamma_0)$. These angles define the orientation of the rotating sample with respect to the laboratory frame (with an initial rotor phase γ_0, which we will henceforth take to be zero, unless otherwise specified). Under these circumstances, the time-dependent irreducible spatial tensor components $R^{\lambda}_{1,m}(t)$ can be written as

$$R^{\lambda}_{1,m}(t) = \sum_{m',p=-1}^{1} D^{(1)}_{m',m}[\Omega_r(t)] D^{(1)}_{p,m'}(\Omega_{PAS})\rho^{\lambda}_{1,p}, \tag{7}$$

by applying the transformation rule of Eq. (6) twice. From Eq. (4) we see that the important spatial tensor components are those with $m = 0$, for which Eq. (7) becomes

$$R^{\lambda}_{1,0}(t) = \sum_{m',p=-1}^{1} D^{(1)}_{m',0}[\Omega_r(t)] D^{(1)}_{p,m'}(\Omega_{PAS})\rho^{\lambda}_{1,p}. \tag{8}$$

The expression for $R^{\lambda}_{1,0}(t)$ is worth considering in some detail. The Euler angles Ω_r contain all of the information needed for describing the net effect of the sample reorientation trajectory on the PAS system of interest, while the products $D^{(1)}_{p,m'}(\Omega_{PAS})\rho^{\lambda}_{1,p}$ in Eq. (8) are invariant. For example, the irreducible spatial components of a chemical shift tensor in its principal axes system $\rho^{\lambda}_{1,p}$ reflects chemical shielding effects associated with local interactions at a given molecular site. Each such component will have an orientation Ω_{PAS} that can be described by $D^{(1)}_{p,m'}(\Omega_{PAS})$ with respect to an intermediate reference frame, which in turn has an orientation Ω_r relative to the laboratory frame, as described by $D^{(1)}_{m',0}[\Omega_r(t)]$. By manipulating the orientation of a bulk sample according to $D^{(l)}_{m',0}[\Omega_r(t)]$, one reorients all individual irreducible spatial tensor components collectively and in concert, so that all $D^{(l)}_{p,m'}(\Omega_{PAS})\rho^{\lambda}_{l,p}$ contributions remain unaffected. As we shall see, judicious choice of a reorientation trajectory $D^{(l)}_{m',0}[\Omega_r(t)]$, for example, rapid spinning of a sample about a fixed angle β, permits broadening effects from anisotropic interactions implicit in $R^{\lambda}_{l,m}$ to be scaled in a predictable way, to zero if desired.

The expression for $R^{\lambda}_{l,0}(t)$ (Eq. 8) is complicated, but simplifies substantially when we consider a particular reorientation trajectory. For rotation about a single axis inclined at an angle β with respect to the magnetic field we have, as noted above,

$$\Omega_r(t) = (0, \beta, \omega_r t). \tag{9}$$

Using this form of Ω_r in the general expression for the rotation matrix elements $D^{(l)}_{j,k}(\alpha, \beta, \gamma)$,

$$D^{(l)}_{j,k}(\alpha, \beta, \gamma) = e^{ij\gamma} d^{(l)}_{j,k}(\beta) e^{ik\alpha}, \tag{10}$$

where $d^{(l)}_{j,k}(\beta)$ is a real combination of trigonometric functions of β, the terms $R_{0,0}$ and $R_{2,0}$ can be rewritten as

$$R_{0,0}(t) = \sum_{m',p=0} e^{im'\omega_r t} d^{(0)}_{m',0}(\beta) D^{(0)}_{p,m'}(\Omega_{PAS})\rho^{\lambda}_{0,p} \tag{11}$$

and

$$R_{2,0}(t) = \sum_{m',p=-2}^{2} e^{im'\omega_r t} d^{(2)}_{m',0}(\beta) D^{(2)}_{p,m'}(\Omega_{PAS})\rho^{\lambda}_{2,p}. \tag{12}$$

The first of these, $R_{0,0}$, is actually independent of time due to the restriction of p and m' to zero. This term thus becomes

$$R_{0,0}(t) = d^{(0)}_{0,0}(\beta) D^{(0)}_{0,0}(\Omega_{PAS})\rho^{\lambda}_{0,0}. \tag{13}$$

Physically, this is expected, because $R_{0,0}$ transforms like a scalar, and thus cannot change during sample reorientation. The $R_{2,0}(t)$ term is more complex, however, and as seen from Eq. (12), leads to a static component accompanied by oscillations in the Hamiltonian (Eq. 4) at frequencies $\pm\omega_r$ and $\pm 2\omega_r$. These terms give rise to spinning sidebands in the frequency domain. The amplitudes of these oscillations in the detected NMR signal depend on $1/\omega_r$, and are thus suppressed at high spinning frequencies. Assuming the fast-spinning limit, we ignore these terms in Eq. (12) and obtain the following simple form for the Hamiltonian:

$$\mathcal{H} \approx -\omega_0 I_z + \omega_\lambda \overbrace{d^{(0)}_{0,0}(\beta)}^{\text{key term}} T^{\lambda}_{0,0} D^{(0)}_{0,0}(\Omega_{PAS})\rho^{\lambda}_{0,0} + \omega_\lambda \overbrace{d^{(2)}_{0,0}(\beta)}^{\text{key term}} T^{\lambda}_{2,0} \sum_{p=-2}^{2} D^{(2)}_{p,0}(\Omega_{PAS})\rho^{\lambda}_{2,p}. \tag{14}$$

The key terms in this equation, which express the angular dependence of the effective Hamiltonian, are directly related to the spherical harmonics Y^m_l

according to

$$d^{(l)}_{m,0}(\beta) = \sqrt{\frac{4\pi}{2l+1}}\, Y_l^m(\beta) e^{-im\phi}. \tag{15}$$

Thus, for $m = 0$, the key terms in Eq. (14) vary simply as Legendre polynomials in $\cos\beta$:

$$d^{(0)}_{0,0}(\beta) = \sqrt{4\pi}\, Y_0^0(\beta) = P_0(\cos\beta) = 1, \tag{16}$$

$$d^{(2)}_{0,0}(\beta) = \sqrt{\frac{4\pi}{5}}\, Y_2^0(\beta) = P_2(\cos\beta) = \frac{1}{2}\left(3\cos^2\beta - 1\right). \tag{17}$$

The first term in ω_λ of Eq. (14) is clearly isotropic, while the second displays the well-known $(3\cos^2\beta - 1)$ dependence observed for all anisotropic first-order interactions (e.g., $\lambda \equiv$ chemical shift anisotropy, dipole–dipole coupling, etc.) under conditions of rapid sample spinning:

$$\mathscr{H} \approx -\omega_0 I_z + \omega_\lambda \rho^\lambda_{0,0} T^\lambda_{0,0} + \omega_\lambda \tfrac{1}{2}(3\cos^2\beta - 1) T^\lambda_{2,0} \sum_{p=-2}^{2} D^{(2)}_{p,0}(\Omega_{PAS}) \rho^\lambda_{2,p}. \tag{18}$$

It is easy to see that selecting a rotation axis of $\beta = 54.74°$, the "magic angle" root of the second-Legendre polynomial $P_2(\cos\beta)$, scales to zero all anisotropic terms that arise from first-order interactions, leaving only the isotropic frequency components remaining. Another way to visualize this spatial dependence is from the perspective of atomic orbital theory: the isotropic parts transform under rotation in the same way as spherically symmetric *s* orbitals, while the anisotropic contributions transform as d_{z^2} orbitals. Though there is no orientation at which the *s* orbital vanishes, the d_{z^2} orbital has a node at $\beta = 54.74°$, as shown in Fig. 1. Seen from the principal axes frame of the first-order interaction, the magnetic field in the MAS experiment traces a path on a cone whose base links the three vertices of a face of an octahedron and at whose center the apex of the cone rests. With the $\mathbf{B}_0$ field directed vertically, rapid and continuous reorientation of the sample about an axis normal to a face of the octahedron corresponds to the Magic-Angle Spinning trajectory (Fig. 1). A discrete version of this same sample reorientation path exists, namely Magic-Angle Hopping (MAH) [16, 17]. In this experiment, the field is hopped sequentially and discretely between the three vertices of a face of the octahedron shown in Figure 1. The MAH technique yields narrow isotropic lines without spinning sidebands, though the method suffers from long 2D measurement times, which have limited its practical utility as compared with MAS.

Dynamic implementation of cubic (octahedral) symmetry is thus the basis by which Magic-Angle Spinning experiments narrow anisotropic spectral features of spin-$\frac{1}{2}$ species in solids. One caveat is that the rate of sample rotation must be high with respect to the strength of the anisotropic first-order interactions. Though this criterion can often be met in the case of chemical shift interactions,

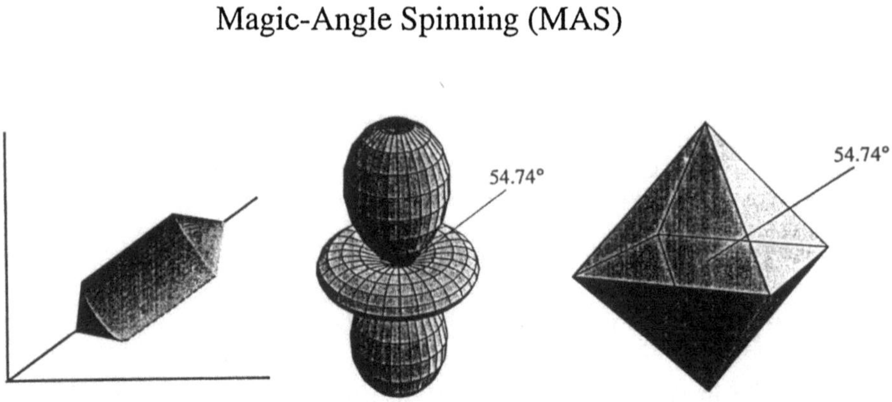

Fig. 1. Magic-Angle Spinning/Hopping averages first-order interactions, which have the same spatial dependence as d_{z^2} atomic orbitals. Averaging of first-order anisotropic broadening effects on NMR spectra can be achieved by rotating a sample about an axis inclined at an angle of $\beta = 54.74°$ relative to the static magnetic field $\mathbf{B}_0$, which corresponds to a path that links the vertices of a face of an octahedron. In Magic-Angle Spinning, the sample trajectory is rapid and continuous, whereas for Magic-Angle Hopping the reorientation is discrete and interrupted. Used with permission from Ref. [5]

it frequently requires supplemental spin decoupling and the use of multiple-pulse sequences to remove additional broadening from stronger dipolar interactions. As is well-known, the isotropic components for the chemical shift (ω_{iso}^{CS}) and scalar coupling (ω_{iso}^{J}) interactions are non-zero, while those for first-order dipole–dipole and first-order quadrupolar coupling interactions (of the central transition) vanish ($\omega_{iso}^{D} = \omega_{iso}^{Q(1)} = 0$).

2.2 Higher-Order Averaging Trajectories

While MAS and multiple-pulse methods yield high-resolution NMR spectra of spin-$\frac{1}{2}$ nuclei, these techniques are only marginally successful for the study of higher-spin systems. The main reason is the quadrupole interaction, which is absent for spin-$\frac{1}{2}$ nuclei. This interaction is frequently strong enough in systems with spin $> \frac{1}{2}$ to render insufficient the first-order description presented in the previous section (beginning with Eq. 4). The practical consequence of the addition of higher-order terms is that the simple transformation properties of the first-order Hamiltonian are lost, and with them the effectiveness of magic-angle spinning. As first described in 1988 by Pines and coworkers at Berkeley and by Llor and Virlet from Saclay [19], more complicated sample reorientation trajectories are required to remove anisotropic broadening influences on the solid-state NMR spectra of quadrupolar nuclei. Subsequent publications from the two groups presented the theory behind removal of higher-order spectral broadenings [18, 33], with the first experimental realizations of approaches for averaging second-order anisotropic effects, namely Double Rotation (DOR) and Dynamic-Angle Spinning (DAS), achieved in Berkeley during this time [33, 39].

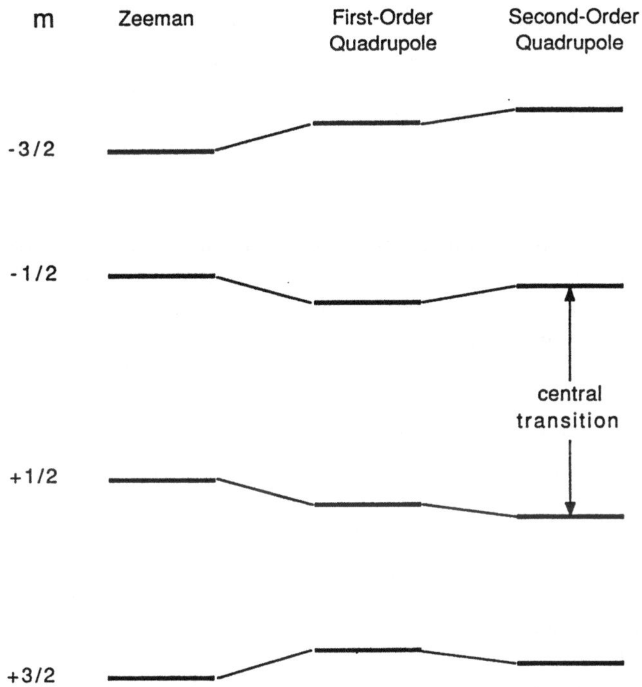

Fig. 2. Diagram illustrating first- and second-order frequency shifts of the Zeeman energy levels of a spin-$\frac{3}{2}$ nucleus (e.g., ^{23}Na) that result from interaction between the nuclear quadrupole moment and local electric field gradients. While the $(+\frac{1}{2} \leftrightarrow -\frac{1}{2})$ central transition is unaffected by first-order effects, the second-order shift can produce a broad distribution of frequencies in a powdered polycrystalline sample. Used with permission from Ref. [21]

Recently, the theoretical foundation of these topics has been presented in considerable detail by Goldman et al. [13], Sun et al. [20], and Emsley and Pines [20a].

Consider, for example, the specific case of a quadrupolar spin $I = \frac{3}{2}$ nucleus (e.g., ^{23}Na) in a solid lattice, whose energy levels are shown in Fig. 2 under the additive influences of a large static magnetic field and first- and second-order quadrupolar interactions. Reading from the left in Fig. 2, the degeneracy of the levels is removed by placement of the sample in a large static $\mathbf{B}_0$ field, which splits the levels according to magnetic quantum number $m = +\frac{3}{2}, +\frac{1}{2}, -\frac{1}{2}$, or $-\frac{3}{2}$ (increasing energy). Chemical shift effects can be accommodated in a straightforward way, both for spin-$\frac{1}{2}$ and spin $> \frac{1}{2}$ systems, as a perturbation of the Zeeman contribution (see Eq. 4 above). Being comparatively small, they have been omitted in this presentation for reasons of simplicity. Strong interactions between the nuclear quadrupole moment and local electric field gradients produce the first- and second-order shifts shown, which often dominate the spectra of quadrupolar nuclei, including ^{11}B, ^{17}O, ^{23}Na, ^{27}Al and indeed over half of the NMR-active isotopes of the Periodic Table. The $(+\frac{1}{2} \leftrightarrow -\frac{1}{2})$ central transition is notably unaffected by first-order quadrupolar interactions, but is shifted by second-order effects [7].

First-order quadrupolar broadening of the satellite transitions, for example the ($\pm\frac{3}{2}\leftrightarrow\pm\frac{1}{2}$) transitions shown in Fig. 2, can however be severe [22], with the result that such spectral features are often difficult to observe except under conditions of fast MAS. Under these circumstances, Satellite Transition Spectroscopy (SATRAS) methods have proven useful for improving spectral resolution, because anisotropic second-order broadening of MAS satellite peaks is reduced compared to that of the central transition. This is particularly helpful in the case of spin-$\frac{5}{2}$ nuclei, such as ^{27}Al, for which the MAS line widths associated with the ($\pm\frac{3}{2}\leftrightarrow\pm\frac{1}{2}$) satellite transitions are only 0.3 that of the central peak (though some anisotropic broadening effects invariably remain). This holds similarly true for the ($\pm\frac{5}{2}\leftrightarrow\pm\frac{3}{2}$) transitions of spin-$\frac{9}{2}$ nuclei, e.g. ^{93}Nb, though in general, enhancement of resolution using SATRAS is less efficient for other transitions or nuclei with spins other than $I=\frac{5}{2}$ or $I=\frac{9}{2}$. Nevertheless, reliable fast MAS probeheads have made the SATRAS technique a convenient means of examining the local structure of a variety of ^{17}O-, ^{23}Na-, ^{27}Al-, or ^{51}V-containing inorganic compounds, including glasses [23–25]. Jäger has recently provided a thorough discussion of SATRAS procedures and applications, and we refer the reader to reference [8] for more details. In this review, we concentrate on techniques for observing the central transition of quadrupolar nuclei with high resolution, specifically establishing experimental strategies for complete averaging of higher-order anisotropic interactions.

We showed in the previous section that finding a strategy for averaging an interaction depends on establishing how that interaction can be modulated, that is, how it is affected by different transformations. In the case of first-order anisotropies, we could write down the effective Hamiltonian (Eq. 4) essentially by inspection, given the commutation properties of the spin degrees of freedom. For second-order (and higher) anisotropies, a more systematic approach is required. To that end, we first summarize the equations of Static Perturbation Theory, and then apply them to the case(s) of particular interest.

As before, we begin by writing the Hamiltonian as $\mathscr{H}=\mathscr{H}_Z+\mathscr{H}_\lambda$, with $\mathscr{H}_\lambda$ again representing the perturbation. The energies and eigenstates of this system are defined as usual by

$$\mathscr{H}|v_m\rangle = E_m|v_m\rangle. \tag{19}$$

Perturbation expansions for these quantities yield

$$E_m = E_m^{(0)} + E_m^{(1)} + E_m^{(2)}\ldots \tag{20}$$

$$|v_m\rangle = |m\rangle + |v_m^{(1)}\rangle + \ldots, \tag{21}$$

where the zeroth-order contributions are

$$\mathscr{H}_Z|m\rangle = E_m^{(0)}|m\rangle. \tag{22}$$

Higher-order corrections to the zeroth-order energies are

$$E_m^{(1)} = \langle m|\mathscr{H}_\lambda|m\rangle, \tag{23}$$

and

$$E_{\rm m}^{(2)} = \sum_{\rm m' \neq m} \frac{\langle m|\mathscr{H}_\lambda|m'\rangle\langle m'|\mathscr{H}_\lambda|m\rangle}{E_{\rm m}^{(0)} - E_{\rm m'}^{(0)}}, \tag{24}$$

and to the eigenstates,

$$|v_{\rm m}^{(1)}\rangle = \sum_{\rm m' \neq m} |m'\rangle \frac{\langle m'|\mathscr{H}_\lambda|m\rangle}{E_{\rm m}^{(0)} - E_{\rm m'}^{(0)}}. \tag{25}$$

These equations can be cast in operator form by viewing the perturbation procedure as a systematic diagonalization of the Hamiltonian accurate to progressively higher orders. We have in fact

$$\mathscr{H} = V\mathscr{H}_{\rm diag}V^\dagger, \tag{26}$$

$$\mathscr{H}_{\rm diag} = H_{\rm Z} + \mathscr{H}_{\rm diag}^{(1)} + \mathscr{H}_{\rm diag}^{(2)} + \ldots, \tag{27}$$

$$V = \mathbf{1} + V^{(1)} + \ldots, \tag{28}$$

with

$$\mathscr{H}_{\rm diag}^{(\rm n)} = \sum_{\rm m} |m\rangle E_{\rm m}^{(\rm n)} \langle m| \tag{29}$$

and

$$V^{(\rm n)} = \sum_{\rm m} |v_{\rm m}^{(\rm n)}\rangle\langle m|. \tag{30}$$

Let us now apply this formalism to the problem of computing the effective Hamiltonian, correct through second order, for a rank-2 interaction in the presence of a strong magnetic field. The two terms in the Hamiltonian are

$$\mathscr{H}_{\rm Z} = -\omega_0 I_{\rm z} \tag{31}$$

and

$$\mathscr{H}_\lambda = \omega_\lambda \sum_{\rm m=-2}^{2} (-1)^{\rm m} R_{2,-\rm m}^\lambda T_{2,\rm m}^\lambda. \tag{32}$$

The zeroth-order states and energies are just those due to the dominant Zeeman interaction:

$$-\omega_0 I_{\rm z}|m\rangle = -m\omega_0|m\rangle. \tag{33}$$

The first-order energy correction is computed from

$$\begin{aligned}\langle m|\mathscr{H}_\lambda|m\rangle &= \langle m|\omega_\lambda \sum_{\rm m'=-2}^{2} (-1)^{\rm m'} R_{2,-\rm m'}^\lambda T_{2,\rm m'}^\lambda|m\rangle \\ &= \omega_\lambda \sum_{\rm m'=-2}^{2} (-1)^{\rm m'} R_{2,-\rm m'}^\lambda \langle m|T_{2,\rm m'}^\lambda|m\rangle. \end{aligned}\tag{34}$$

The matrix element can be simplified by noting one of the selection rules for spherical tensor operators:

$$\langle m|T_{l,m'}|m''\rangle \propto \delta_{m,m'+m''}. \tag{35}$$

Using this result in Eq. (34) shows that only the $T^{\lambda}_{2,0}$ term contributes. Converting the first-order energies into an operator, as described in Eq. (29), yields $\mathscr{H}^{(1)}_{\text{diag}} = \omega_\lambda R^{\lambda}_{2,0} T^{\lambda}_{2,0}$, and thus an effective Hamiltonian identical to what we found in the previous section.

Having confirmed the first-order calculation, we now calculate the second-order term. (Henceforth the λ designation associated with $R^{\lambda}_{l,m}$ and $T^{\lambda}_{l,m}$ will be assumed.) Substituting $\mathscr{H}_\lambda$ in the form of Eq. (32) into Eq. (24), we find

$$E^{(2)}_m = \frac{\omega^2_\lambda}{\omega_0} \sum_{k \neq m} \frac{\langle m|\sum_{m'}(-1)^{m'} R_{2,-m'} T_{2,m'}|k\rangle\langle k|\sum_{m''}(-1)^{m''} R_{2,-m''} T_{2,m''}|m\rangle}{k-m}. \tag{36}$$

Rearranging to separate the matrix elements from the spatial tensors gives

$$E^{(2)}_m = \frac{\omega^2_\lambda}{\omega_0} \sum_{m',m''} (-1)^{m'+m''} R_{2,-m'} R_{2,-m''} \sum_{k \neq m} \frac{\langle m|T_{2,m'}|k\rangle\langle k|T_{2,m''}|m\rangle}{k-m}. \tag{37}$$

Application of the selection rule in Equation 35 shows that the only values of k that contribute to the second sum are those for which both $k = m - m'$ and $k = m + m''$. Taken together these conditions require that $m' + m'' = 0$. We can thus drop m'' in favor of m', and simplify the energy correction to

$$E^{(2)}_m = \frac{\omega^2_\lambda}{\omega_0} \sum_{m' \neq 0} R_{2,-m'} R_{2,m'} \frac{\langle m|T_{2,m'}|m-m'\rangle\langle m-m'|T_{2,-m'}|m\rangle}{-m'}. \tag{38}$$

Finally, making the substitution

$$|m-m'\rangle\langle m-m'| = \mathbf{1} - \sum_{l \neq m-m'} |l\rangle\langle l| \tag{39}$$

yields

$$E^{(2)}_m = -\frac{\omega^2_\lambda}{\omega_0} \sum_{m' \neq 0} R_{2,-m'} R_{2,m'} \frac{\langle m|T_{2,m'} T_{2,-m'}|m\rangle}{m'}. \tag{40}$$

These energies are easily converted to the operator $\mathscr{H}^{(2)}_{\text{diag}}$, which, after a last rearrangement, yields

$$\mathscr{H}^{(2)}_{\text{diag}} = \frac{\omega^2_\lambda}{\omega_0} \sum_{m' > 0} R_{2,-m'} R_{2,m'} \frac{[T_{2,-m'}, T_{2,m'}]}{m'}. \tag{41}$$

Equation 41 is a compact representation of the second-order term in the Hamiltonian, but we have not yet addressed the question of how it transforms. We will consider the spin and space degrees of freedom separately. This procedure is not entirely accurate, because at magnetic fields weak enough to require a second-order description of the quadrupole interaction, these two sets of variables are no longer independent. In other words, truncation by the magnetic field is not complete. The mathematical expression of this fact is that, to be consistent with the use of second-order energies, we should use the first-order correction to the eigenstates, which in this case takes the form

$$V^{(1)} = \frac{\omega_\lambda}{\omega_0} \sum_{m \neq 0} (-1)^m \frac{R_{2,-m} T_{2,m}}{m}. \tag{42}$$

This term tilts the eigenstates away from the Zeeman states, but the tilting is small ($\sim \omega_\lambda/\omega_0$). Ignoring this term, thus, leads to slight errors in the intensities that in practical cases are negligible. A more detailed discussion is contained in Ref. [13].

Continuing then, we first consider the spin variables, $[T_{2,-m'}, T_{2,m'}]$. Products of spherical tensors are in general reducible, the reduction being expressible in terms of Clebsch–Gordan coefficients:

$$T_{l_1,m_1} T_{l_2,m_2} = \sum_{L,M} \langle l_1, m_1, l_2, m_2 | l_1, l_2, L, M \rangle T_{L,M}. \tag{43}$$

Here, the $T_{L,M}$ are new irreducible spherical tensors of the spin operators, and the weights are Clebsch–Gordan coefficients. Applying this expansion to the terms in $[T_{2,-m'}, T_{2,m'}]$ shows first that only $T_{L,M}$ with $0 \leqq L \leqq 4$ and $M = 0$ are possible. Next, the symmetry of the Clebsch–Gordan coefficients with respect to interchange of l_1, m_1 with l_2, m_2 is given by the factor $(-1)^{l_1+l_2+L}$. Thus, in the commutator $[T_{2,-m'}, T_{2,m'}]$, terms with *even* L vanish, while terms with *odd* L survive. The end result is that in the second-order term there are only spin variables transforming as the $M = 0$ component of rank-1 and rank-3 tensors. Now, the $M = 0$ component of a rank-1 tensor (in spin space) is just I_z. The rank-3 term in general is complicated [3], since it contains terms like I_z^3, but we are particularly concerned with the central transition, for which $I_z^3 \sim I_z$. Thus, the spin part of the second-order Hamiltonian acts, for the central transition at least, like I_z for a two-level system. One important consequence of this is that the overall spin contribution to the Hamiltonian varies as I_z, precluding the use of multiple-pulse strategies alone for averaging anisotropic second-order interactions without annihilating the entire Hamiltonian. Nevertheless, the I_z-dependence of the second-order effects opens the way to the simple use of spin echoes in a variety of line-narrowing experiments that rely on reorientation of the sample, as we describe in Sect. 4.

Next, let us consider how the spatial degrees of freedom, $R_{2,-m'} R_{2,m'}$, transform. Recall that each spatial tensor R expresses the principal components of an interaction tensor in the laboratory frame through (see Eq. 6)

$$R_{l,m} = \sum_k D^{(l)}_{k,m}(\Omega) \rho^\lambda_{l,k}. \tag{44}$$

As a consequence, the product of two spatial tensors can be rewritten as

$$R_{2,-m'} R_{2,m'} = \sum_{j,k} D^{(2)}_{j,-m'}(\Omega) D^{(2)}_{k,m'}(\Omega) \rho^\lambda_{2,j} \rho^\lambda_{2,k}. \tag{45}$$

As in the case of the spin variables, the product of the rank-2 R tensors can lead only to terms transforming as rank-0 through rank-4. The product of

Wigner rotation matrix elements can be rewritten with Clebsch–Gordan coefficients to obtain [15]

$$D^{(2)}_{j,-m'}(\Omega)D^{(2)}_{k,m'}(\Omega)$$
$$= \sum_{L=0}^{4} \sum_{M,N=-L}^{L} \langle 2,j,2,k|2,2,L,M\rangle\langle 2,-m',2,m'|2,2,L,N\rangle D^{(L)}_{M,N}(\Omega). \tag{46}$$

The second Clebsch–Gordan coefficient permits only $N = 0$ terms to contribute to the sum. The first coefficient ensures that, once we sum over j and k, all odd orders will vanish, simplifying significantly the resulting expresson by reducing the number of contributing terms. This is again due to the symmetry with respect to interchange of j and k in this term: symmetric, when $2 + 2 + L$ is even, and antisymmetric, when $2 + 2 + L$ is odd. Thus, only contributions for which $L = 0$, 2, or 4 prevail, so that we obtain a spatial term with components transforming like rank-0, rank-2, and rank-4 tensors. For the central transition, therefore, the effective second-order term in the Hamiltonian takes the form

$$\mathscr{H}^{(2)}_{\text{diag}} = \frac{\omega_\lambda^2}{\omega_0}\left[a_0 R_{0,0} + a_2 R_{2,0} + a_4 R_{4,0}\right] I_z, \tag{47}$$

where the $R_{L,0}$ are the irreducible spherical tensors of the spatial operators, the a_i are constants, and all other additive constants that exert no influence on the dynamics of the spin system under conditions of sample reorientation have been dropped to simplify the presentation.

Let us now consider the effect of spatial modulation on this Hamiltonian, in particular, of spinning the sample about some axis at an angle β with respect to the magnetic field. First, note that the discussion leading to Eq. (12) is quite general. In the present case, we must consider time-dependent variation of the spatial tensors $R_{L,0}$ with $L = 0, 2, 4$, and so, for rotation of the sample about a fixed axis, we have

$$R_{L,0}(t) = \sum_{m',p=-L}^{L} e^{im'\omega_r t} d^{(L)}_{m',0}(\beta) D^{(L)}_{p,m'}(\Omega_{\text{PAS}}) \rho^{\lambda}_{L,p}. \tag{48}$$

Here, the $\rho^{\lambda}_{L,p}$ are the spherical tensor components of the products $\rho^{\lambda}_{2,j}\rho^{\lambda}_{2,k}$, which have been computed using the formalism of Eq. (43). Upon ignoring oscillating terms due to the rotation (using the same argument that led to Eq. 14), the second-order Hamiltonian takes the form

$$\mathscr{H}^{(2)}_{\text{diag}} = \frac{\omega_\lambda^2}{\omega_0}\Bigg[a_0 \overbrace{d^{(0)}_{0,0}(\beta)}^{\text{key term}} D^{(0)}_{0,0}(\Omega_{\text{PAS}})\rho^{\lambda}_{0,0} + a_2 \overbrace{d^{(2)}_{0,0}(\beta)}^{\text{key term}} \sum_{p=-2}^{2} D^{(2)}_{p,0}(\Omega_{\text{PAS}})\rho^{\lambda}_{2,p}$$
$$+ a_4 \overbrace{d^{(4)}_{0,0}(\beta)}^{\text{key term}} \sum_{p=-4}^{4} D^{(4)}_{p,0}(\Omega_{\text{PAS}})\rho^{\lambda}_{4,p}\Bigg] I_z. \tag{49}$$

The factors $d^{(0)}_{0,0}(\beta)$ and $d^{(2)}_{0,0}(\beta)$ have already appeared in Eqs. (16) and (17) for the first-order case; the new term $d^{(4)}_{0,0}(\beta)$ is similarly related directly to the spherical harmonics Y_l^m, taking the form

$$d_{0,0}^{(4)}(\beta) = \sqrt{\frac{4\pi}{9}}\, Y_4^0(\beta) = P_4(\cos\beta) = \frac{1}{8}\left(35\cos^4\beta - 30\cos^2\beta + 3\right). \quad (50)$$

In common with the first-order Hamiltonian in Eq. (14), the second-order Hamiltonian in Eq. (49) has rank-0 and rank-2 anisotropies. However unlike the first-order expression, the second-order Hamiltonian possesses a rank-4 term, as well. The rank-0 term is a scalar, and implies that there will be a frequency shift, even after all broadening influences are removed. From Eq. (41), we can expect this shift to vary as $\omega_\lambda^2/\omega_0$ for higher-order interactions, thus increasing at low fields and diminishing in high fields. To obtain high-resolution NMR spectra in the presence of such higher-order effects, it will generally be necessary to average both the rank-2 and rank-4 terms, which can be done using either discrete or continuous trajectories. Each of these terms can be separately averaged, in the continuous case, by spinning the sample rapidly around the proper axis inclined at an angle β, such that $d_{0,0}^{(L)}(\beta) = P_L(\cos\beta) = 0$. In the discrete case, the sample must spend equal times at a minimum of $L+1$ orientations distributed evenly around this axis. For $L = 2$ the angle is the usual magic angle, and these two experiments are just Magic-Angle Spinning and Magic-Angle Hopping, respectively, as described in Sect. 2.1. Clearly, one could perform $L = 4$ analogs of these experiments, but unfortunately, the magic angles for the two cases are not the same: the rank-2 magic angle is 54.74°, while rank-4 magic angles are 30.56° and 70.12°. This is just a mathematical statement of the well-known fact that spinning a sample around any one axis does not remove all of the second-order quadrupolar anisotropy. Figure 3 shows plots of

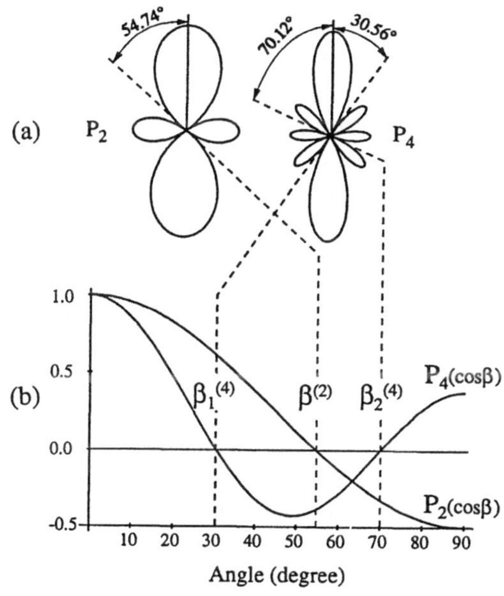

Fig. 3a, b. Under conditions of rapid sample spinning, anisotropic first- and second-order interactions contain terms that scale as the second- and fourth-Legendre polynomials $P_2(\cos\beta_i)$ and $P_4(\cos\beta_i)$ (Eqs. 17 and 50). Plots of P_2 and P_4 in **a** polar coordinates and in **b** Cartesian coordinates, with the roots of the respective functions indicated by the *dashed lines*. Used with permission from Ref. [20]

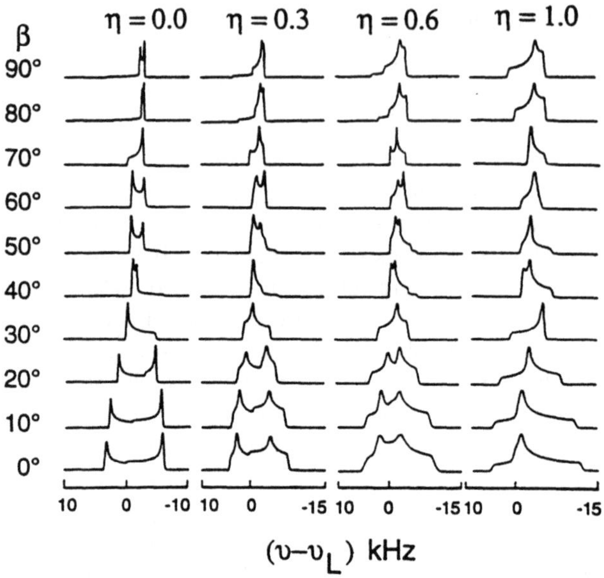

Fig. 4. Variation of powder line shapes arising from second-order quadrupolar interactions simulated for a spin-$\frac{3}{2}$ system under conditions of rapid sample spinning about a single axis fixed at different angles β with respect to the static $\mathbf{B}_0$ field. The spinning speed has been assumed to be much larger than the amplitude of the second-order quadrupolar interaction, so that no spinning sidebands are generated. The effect of the asymmetry parameter η on the spectral features is also shown. Used with permission from Ref. [27]

$P_2(\cos\beta)$ and $P_4(\cos\beta)$ in polar and Cartesian representations, as functions of the angle formed between the rotation axis of a spinning sample and the applied $\mathbf{B}_0$ field.

While complete suppression of the anisotropy is not possible by spinning about a single axis, partial line narrowing can still be achieved [26–31], the extent to which varies according to the strength and asymmetry of the higher-order interaction and the angle formed by the rotation axis and the applied field. This is shown in Fig. 4, where powder line shapes arising from second-order quadrupolar interactions have been simulated for spin-$\frac{3}{2}$ systems (e.g., ^{23}Na) with differing asymmetry parameters η and under conditions of fast spinning about a single axis inclined at different angles β with respect to the static field [27]. At an angle of $\beta = 0°$ relative to $\mathbf{B}_0$, the full second-order powder pattern is obtained. For $\beta = 54.74°$, the magic angle that scales P_2 to zero, the broadening is due solely to the contribution from the P_4 term. The technique of Variable-Angle Spinning (VAS) has been used to improve resolution over that achievable by MAS, through selection of an angle formed by the spinning axis and $\mathbf{B}_0$ to minimize (but not eliminate) the sum of the P_2 and P_4 anisotropic contributions [27, 28, 31]. Though not eliminating anisotropic broadening effects completely, VAS can nevertheless yield line shapes that are narrower than MAS through judicious selection of a spinning angle that reduces the net effects of the ansiotropy on spectral features. Line widths of several kilohertz, however, may

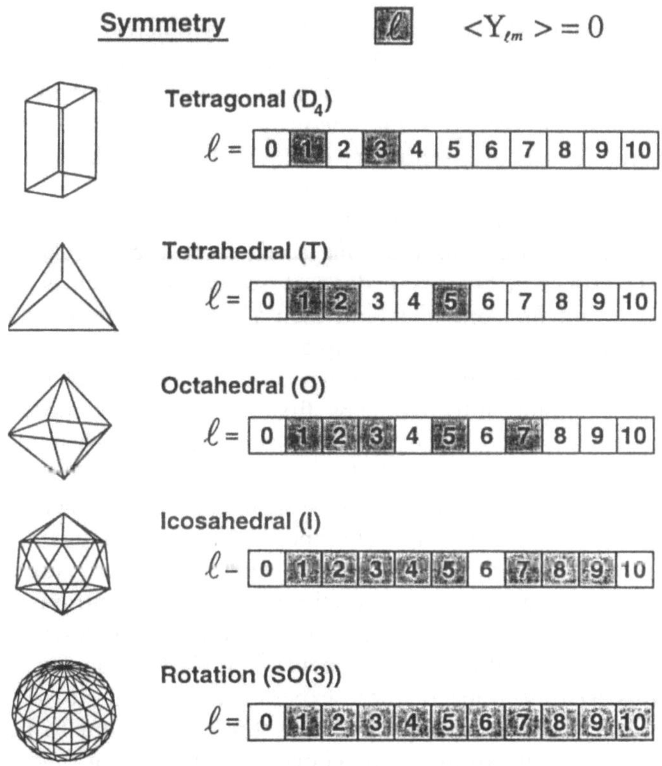

Fig. 5. Averaging of spherical harmonics Y_l^m of different ranks under point subgroups of the full rotation group $SO(3)$. The *shaded numbers* indicate those ranks l for which the average $\langle Y_l^m \rangle = 0$. Whereas rank $l = 2$ is averaged to zero by trajectories with cubic (i.e., octahedral) symmetry, ranks $l = 2$ and $l = 4$ are averaged simultaneously to zero only for trajectories possessing at least icosahedral symmetry. Isotropic reorientation processes, such as occur in low viscosity liquids, correspond to the full symmetry of $SO(3)$, which averages all ranks $l \neq 0$. Used with permission from Ref. [5]

still result, complicating the analysis when overlapping features from multiple sites are confronted.

Given that rotation around a single axis is not sufficient to remove the second-order anisotropies, what sort of trajectory should be sought? We mentioned in the previous section that Magic-Angle Spinning imposes a trajectory with octahedral symmetry on the sample. More generally, Samoson, Sun, and Pines have shown in detail how to determine the symmetry sufficient to average any given interaction to zero [32]. They emphasize finding how the representations of the group of rotations, $SO(3)$, reduce when the lower symmetry of some particular sample reorientation trajectory is imposed. Figure 5 shows how spherical harmonics Y_l^m of different ranks are averaged by trajectories with various symmetries corresponding to subgroups of the full rotation group $SO(3)$. Since the tensors of interest transform like the representations of $SO(3)$, the averaging process will be successful (that is, will yield zero signal for that particular tensor) if the reduction *does not* include the totally symmetric representation. For example, trajectories with octahedral symmetry, such as MAS, average tensors of ranks 1–3, 5, and 7, but not 4 or 6. While rank-2 tensors are averaged by octahedral sample reorientation paths, tetrahedral symmetry is in fact sufficient. To average simultaneously tensors of rank-2 *and* tensors of rank-4, however, icosahedral symmetry is required, with such a path sufficient to remove tensors of ranks 1–5 and 7–9 as well.

There are several possible schemes for implementing trajectories with icosahedral symmetry. One is purely discrete: one could force the sample to hop among six distinct vertices of an icosahedron. A second approach is to spin the sample rapidly about first one axis, and then a second, chosen such that a subset of the icosahedral vertices are sampled during the first period, and the others during the second. Finally, one can take the rather different approach of splitting the averaging into two steps: first, a trajectory is chosen to average one of the tensors (rank-2, for example) to zero, and at each point of that path, a second trajectory is imposed that averages the other tensor (rank-4, for example) to zero. In fact, all three of these experimental schemes have been implemented in some form. The discrete experiment is called Dynamic Angle Hopping; the partially continuous version is called Dynamic-Angle Spinning; and the iterative averaging scheme, practiced as a purely continuous trajectory, is called Double Rotation. In the succeeding sections we will describe the experimental implementations of these experiments, and show examples of their application. At this writing, none of these experiments are easy to do. Therefore, before describing them in detail, we review the information to be gained from such studies, as justification for the effort of implementing them.

After averaging away the anisotropies, a sharp resonance at shift δ_{obs} may be obtained for each resolvable site. This shift includes contributions from both isotropic first-order (e.g., chemical shift, $\delta_{\text{iso}}^{(\text{CS})}$) and isotropic higher-order (e.g., second-order quadrupolar, $\delta_{\text{iso}}^{(\text{Q},2)}$) effects:

$$\delta_{\text{obs}} = \delta_{\text{iso}}^{(\text{CS})} + \delta_{\text{iso}}^{(\text{Q},2)}. \tag{51}$$

While the isotropic chemical shift $\delta_{\text{iso}}^{(\text{CS})}$ in ppm is independent of the magnetic field, the isotropic second-order quadrupolar shift $\nu_{\text{iso}}^{(\text{Q},2)}$ for the central transition of a spin I species scales (in Hertz) inversely with the Larmor frequency $\omega_0 = 2\pi\nu_0$:

$$\nu_{\text{iso}}^{(\text{Q},2)} = -\frac{3}{40}\frac{C_{\text{Q}}^2[I(I+1)-\frac{3}{4}]}{\nu_0 I^2(2I-1)^2}\left(1+\frac{\eta^2}{3}\right), \tag{52}$$

where η is the asymmetry parameter of the electric field gradient and the quadrupolar coupling constant C_{Q} is defined by $C_{\text{Q}} \equiv e^2qQ/h$. The quadrupolar coupling constant C_{Q} is comprised of eq, the z-component of the electric field gradient at the nucleus, eQ, the nuclear electric quadrupole moment, and Planck's constant h. The shift $\nu_{\text{iso}}^{(\text{Q},2)}$, and indeed all effects of the quadrupole interaction on the central transition, vanish in the limit of very high field. However, in many cases of interest, this would mean fields well in excess of 10 T and very fast spinning, so this approach is not widely used (though see [29, 30]).

Because of differing field dependences, the contributions from $\delta_{\text{iso}}^{(\text{CS})}$ and $\delta_{\text{iso}}^{(\text{Q},2)}$ to the observed peak position can be distinguished by making measurements at different $\mathbf{B}_0$ field strengths (or by correlation with anisotropic line shapes acquired from separate MAS, DAS, or VAS studies.) These important experimental parameters contain complementary information on local chemical structure and are the primary target of high-resolution solid-state NMR studies on quadrupolar systems.

To summarize, the isotropic parameters of first- and higher-order interactions will be obscured in solids by anisotropic broadening effects, which can be removed by creative manipulation of the spin or spatial tensors using multiple-pulse or sample reorientation techniques. Equation (41) provides the theoretical foundation for the design of higher-order averaging schemes, which must be implemented to obtain high-resolution solid-state NMR spectra of quadrupolar systems. The key observation is that there exists no single rotation axis about which a sample may be rapidly spun that will remove completely higher-order anisotropic broadening influences on an NMR spectrum. However, by allowing for more complicated (e.g., multiple angle) trajectories, these limitations can be overcome, and we turn our attention now to several ways in which higher-order averaging may be experimentally implemented.

3 Double Rotation

3.1 DOR Implementation and Probehead Designs

As discussed in the previous section, to eliminate spectral broadening effects associated with second-order interactions it is necessary to direct a sample through a trajectory that averages to zero anisotropies transforming as rank-2 and rank-4 tensors (Eq. 49). We noted that this can be accomplished by using a trajectory that averages the rank-2 term and, at each point of this trajectory, implementing a secondary trajectory that averages the rank-4 term. The continuous version of this strategy involves spinning the sample about a *time-dependent* axis, achieved by spinning the sample at a rank-4 magic angle (typically 30.56°) with respect to the axis of a rotor that is itself spinning at the rank-2 magic angle (54.74°) with respect to the magnetic field. This experiment is called Double Rotation (DOR), and was first demonstrated by Samoson et al. [33, 34].

Figure 6a contains a schematic diagram illustrating implementation of the Double Rotation experiment. A small inner rotor containing the powdered sample spins at an angle inclined 30.56° with respect to the axis of a larger outer rotor, which itself rotates at an angle of 54.74° with respect to the static $\mathbf{B}_0$ field. Practically, this has been implemented by nesting the inner rotor inside the outer rotor, as shown in the exploded diagram in Fig. 6b, and spinning both rotors pneumatically with separate control of bearing and drive air for each

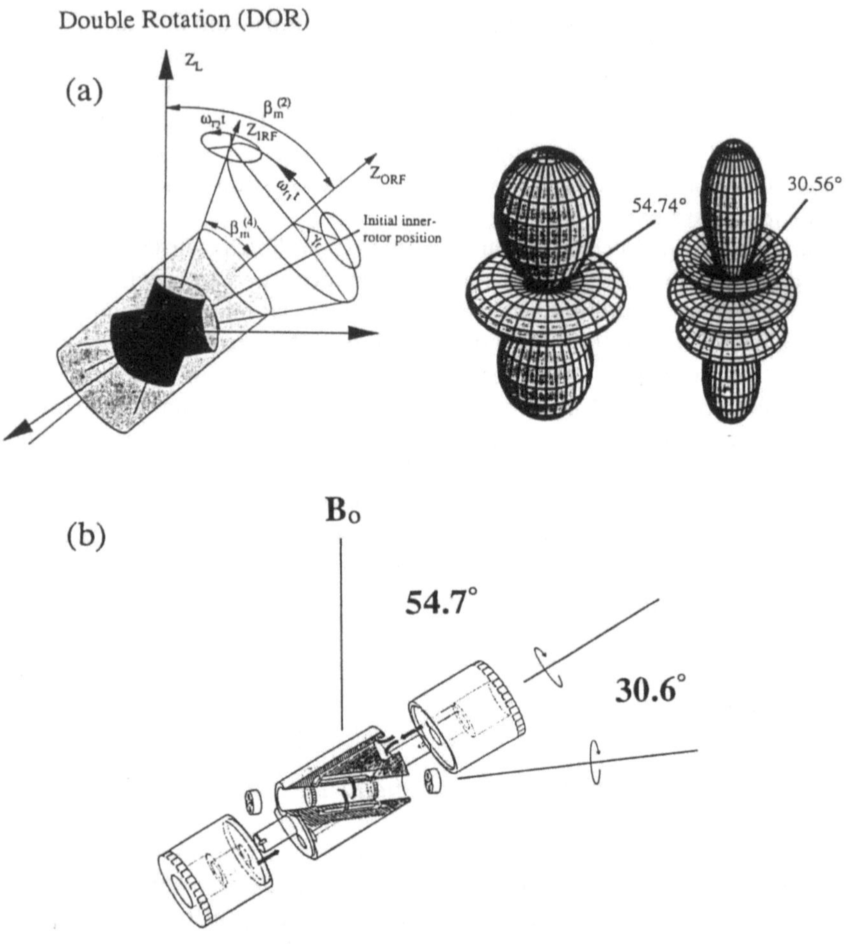

Fig. 6a. Schematic diagram showing the relative orientations and rotation frequencies of the inner and outer spinners used in the Double Rotation experiment. As the outer rotor spins about an angle inclined 54.74° with respect to $\mathbf{B}_0$, the magic angle corresponding to the root of $P_2(\cos\beta)$, an inner rotor tilted 30.56° relative to the axis of the outer rotor, a magic angle of $P_4(\cos\beta)$, sweeps through a trajectory that averages first- and second-order effects to zero. The spatial dependence of first-order (second-rank) effects resemble *d* orbitals, while second-order (fourth-rank) interactions can be considered a linear combination of *d* and *g* orbitals. The *d* orbitals display a node at 54. 74°, while the *g* orbitals vanish at 30.56° or 70.12°. Adapted with permission from Refs. [5, 20]. **b** Exploded view of the double rotor assembly showing the relative positions of the inner and outer DOR spinners. For the design shown, an air-driven inner rotor containing ≈50 mg of sample spins at ca. 5 kHz, while the outer rotor spins within a solenoid coil and a stator (not shown) at ca. 1 kHz. Adapted with permission from Ref. [35]

[33, 34]. The design shown in Fig. 6b is that of Wu et al. [35], who engineered several improvements to the double rotor device that permitted faster, more stable, simultaneous operation of the two rotors. Recently several commercial designs [36, 37, 38] have appeared on the market, which have produced additional improvements in DOR probehead technology and provided for increased proliferation of the technique. To date, these designs have retained

general features of the original development, for example, the particular combination of DOR rotor angles, which provides maximum NMR sensitivity within inherent geometric and filling factor constraints. Existing probehead designs have also relied on a large solenoid rf coil (not shown in Fig. 6b), which surrounds the outer rotor coaxially. Careful attention to bearing design, moments of rotor inertia, and relative rotor spinning speeds allows the double rotor assembly to be operated in a low torque condition, which provides reduced bearing loads and enhanced spinning stability [35]. Balancing these delicate features, together with the complicated fluid dynamics associated with the bearing and drive air-flows, requires that fine design tolerances be met and maintained. For practical reasons of machinability and robustness, current probehead designs contain dissimilar polymer and ceramic components, which have demonstrated reliable performance. Different material responses to changes in temperature, however, have limited the versatility of existing devices, restricting current DOR investigations to room temperature applications.

Initial development of the Double Rotation technique [33] was accompanied by demonstrations of its utility on quadrupolar ^{23}Na (spin-$\frac{3}{2}$) and ^{17}O (spin-$\frac{5}{2}$) species in simple, polycrystalline systems [33, 35, 39]. More recent investigations have extended DOR to structurally more complicated inorganic solids, including glasses, minerals, and molecular sieves. As shown in Fig. 7, MAS is capable of narrowing the resonances somewhat from the broad static ^{17}O line shape of the silicate mineral wollastonite, but remaining contributions from $P_4(\cos\beta)$ terms produce significant remnant broadening, which yields poorly resolved spectral features. Because of these residual anisotropic influences, faster MAS spinning speeds do not enhance spectral resolution, making structural characterization by MAS alone essentially impossible. However, under conditions of double rotation, the $P_2(\cos\beta)$ and $P_4(\cos\beta)$ contributions (Eq. 49) to the resonance signals are averaged away, providing a high resolution wollastonite ^{17}O DOR spectrum. In this example, eight of the nine distinct oxygen sites in this complicated mineral structure are resolved, with isotropic DOR lines (approximately 70 Hz wide) indicated by the arrows among the manifold of spinning sidebands. The sideband peaks are displaced from their corresponding isotropic center band by integer multiples of the outer rotor spinning frequency and can be distinguished from the isotropic peaks by performing DOR experiments at two different speeds. Spinning sidebands arising from the motion of the inner rotor are difficult to observe, because of the much faster inner rotor spinning speed, which acts to decouple adiabatically the NMR effects of the two rotors [35]. The six well-resolved peaks in the 75–100 ppm shift range in Fig. 7 are from six non-bridging oxygen sites in the wollastonite structure, which have relatively small quadrupolar coupling constants ($C_Q \approx 3$ MHz). By comparison, the peaks at 21 and 28 ppm correspond to three less symmetric bridging oxygen sites (two of whose signals overlap at 28 ppm), which have quadrupolar coupling constants near 5 MHz. The larger values for C_Q for the bridging oxygen atoms in wollastonite complicate the analysis by producing much broader manifolds of spinning sidebands over which the intensities of these signals are spread.

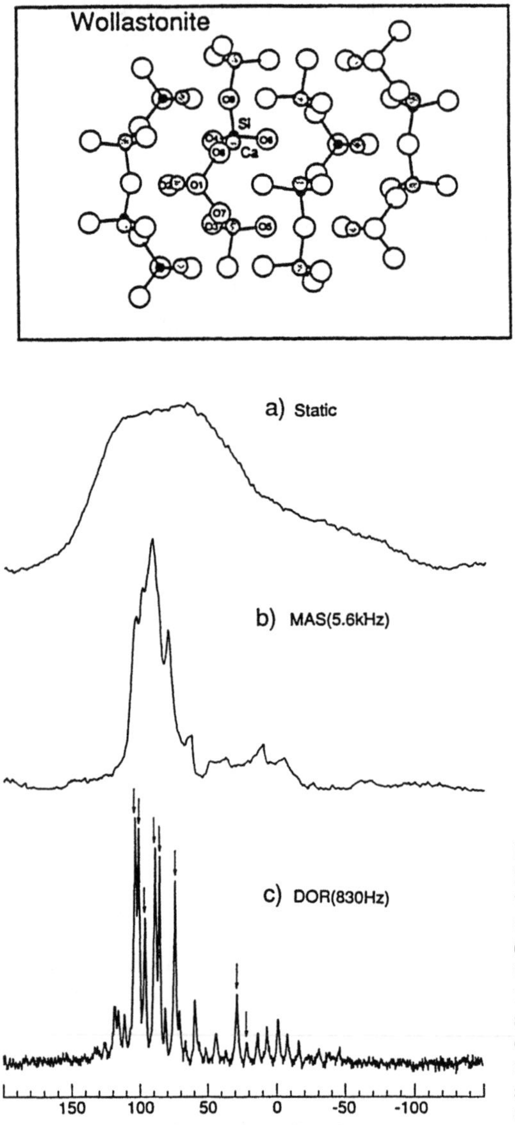

Fig. 7a–c. ^{17}O NMR spectra acquired at 9.4 Tesla of the central transition of a 41 wt% ^{17}O-enriched polycrystalline sample of the silicate mineral wollastonite ($CaSiO_3$) (Structural formula in *box*). **a** Static spectrum; **b** MAS spectrum (spinning speed = 5.6 kHz); **c** DOR spectrum (outer rotor spinning speed = 830 Hz). The *arrows* indicate lines that are resolved from eight of the nine crystallographically distinct oxygen sites in this material. Shifts are referenced to $H_2^{17}O$. Adapted with permission from Refs. [35, 47]

3.2 Synchronized DOR

The resolution difficulties posed by broad or overlapping manifolds of spinning sidebands represent a limiting factor in the use of DOR for structural characterization. The sidebands arise primarily from finite outer rotor spinning speeds, which are on the order of the second-order quadrupolar interactions (after scaling by the motion of the inner rotor) and therefore insufficient to average the oscillating time-dependent terms in Eq. (48) to zero. Nevertheless, advantage can be taken of the inversion symmetry of the relative phase of the inner rotor with respect to the outer rotor and $\mathbf{B}_0$ to simplify DOR spectra through synchronous application of rf pulses. The basis for such simplification lies in the dependence of the spin system's evolution on the relative position of the four different axes shown in Fig. 6a. Assuming the principal-axes-system to be fixed with respect to the inner rotor, evolution of the spin system depends upon the relative positions of the inner rotor, outer rotor, and $\mathbf{B}_0$ axes at the time the rf pulse is applied. Importantly, there are two cases when the three axes are coplanar, which are distinguished by their relative orientations: (1) the outer rotor axis positioned between the $\mathbf{B}_0$ and inner rotor axes (defined as $\gamma_0 = 0°$) and (2) the inner rotor axis positioned between the $\mathbf{B}_0$ and outer rotor axes (defined as $\gamma_0 = 180°$).

As demonstrated by Samoson and Lippmaa [40] and Wu et al. [35], synchronizing the rf pulses with the motion of the outer rotor alternately at $\gamma_0 = 0°$ and $\gamma_0 = 180°$ produces cancellation of all odd spinning sidebands, leaving only even contributions to the manifold and enhancing the intensity of the isotropic centerband. This results in a DOR spectrum that contains side-

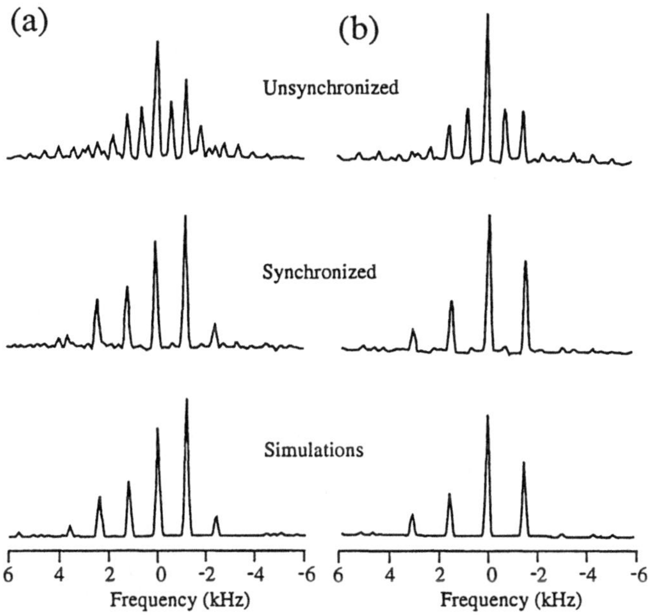

Fig. 8a, b. Effects of synchronizing the radiofrequency excitation pulse with the phase of the inner rotor in ^{23}Na DOR experiments on polycrystalline sodium oxalate (NaC_2O_4) at 9.4 Tesla. Adding spectra obtained at $\gamma_0 = 0°$ and $\gamma_0 = 180°$ (see Fig. 6a) suppresses the odd sidebands, significantly increasing the resolution of the resultant spectrum. Outer rotor spinning speeds of **a** 604 Hz and **b** 800 Hz were used. The computer simulations employed a quadrupolar coupling constant of 405 kHz and an asymmetry parameter of 0.72 for the single sodium site at a Larmor frequency of 105.8 MHz. Used with permission from Ref. [20]

bands at intervals of twice the outer rotor frequency, so that the sideband manifold has the appearance of a spectrum for double the rate of outer rotor rotation. Sun et al. [20] have performed a thorough analysis of sideband patterns under conditions of double rotation (and also dynamic-angle spinning) and shown that it is not possible to cancel all sidebands without canceling the entire signal. Nevertheless, appreciable simplification of DOR spectra can be achieved through suppression of half of the spinning sidebands, as shown for example in the synchronized ^{23}Na DOR spectrum of polycrystalline sodium oxalate (NaC_2O_4) in Fig. 8. Samoson has, furthermore, shown that suppression of higher-order spinning sidebands might be achieved by incorporating inversion pulses into a synchronized DOR experiment [41]. Recent theoretical and experimental treatments of DOR sidebands by Cochon and Amoureux have also appeared in the literature [42, 43].

3.3 Double Resonance DOR Experiments

Improvements in resolution and sensitivity provided by reorientation of a quadrupolar sample about the two DOR axes can be further augmented in combination with high-power proton decoupling and cross-polarization. This, of course, relies on a double-tuned resonance circuit, which may be non-trivial at high magnetic fields, given the constraint of the large diameter coil surrounding the outer rotor (Fig. 6b). Wu et al. have performed ^{27}Al Double Rotation experiments using a double resonance DOR probehead tuned to ^{27}Al and ^{1}H to improve resolution by decoupling protons or to achieve selective signal enhancements via cross-polarization [44]. As shown in Fig. 9a,b, proton decoupling during acquisition greatly enhances the resolution of the ^{27}Al DOR spectrum for the hydrated aluminophosphate $AlPO_4$-11 over that achievable using single-resonance conditions, resolving isotropic peaks from the five distinct aluminium sites in this material. To achieve comparable resolution in the absence of decoupling, Fig. 9c shows that it is necessary to spin the outer rotor nearly twice as fast, near the technical limit of the DOR probehead. In implementing cross-polarization together with DOR, it is necessary to balance simultaneous demands for slow spinning, which is needed to preserve polarization transfer efficiency and spin locking, and fast spinning, which is required to overcome second-order quadrupolar effects and dipolar interactions between the quadrupolar nuclei. Figure 9d shows the CPDOR spectrum for hydrated $AlPO_4$-11 obtained by spinning the outer rotor at 550 Hz, with proton decoupling during acquisition. Under conditions of CPDOR, selective enhancement of ^{27}Al signals indicates strong coupling between certain aluminum sites and protons associated with adsorbed water. This is particularly evident for the ^{27}Al peak at -29.2 ppm in Fig. 9d, consistent with its resonance assignment to six-coordinate aluminum bound strongly to two adsorbed water molecules. In this way, CPDOR can provide detailed site-specific information on molecular adsorption and structure in these complicated materials.

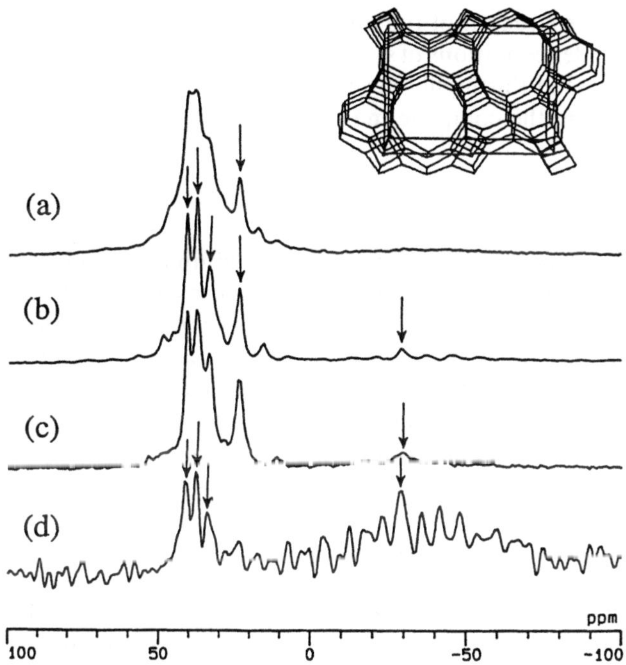

Fig. 9a–d. ^{27}Al DOR spectra of the aluminophosphate molecular sieve $AlPO_4$-11 acquired at 7 T: **a** with the outer rotor spinning at $\omega_0 = 620$ Hz without proton decoupling; **b** at $\omega_0 = 620$ Hz with proton decoupling; **c** at $\omega_o = 1000$ Hz without proton decoupling; and **d** at $\omega_0 = 550$ Hz under conditions of cross-polarization with a 200-μs contact time (CPDOR). Five distinct aluminum sites are resolved in **b** (see *arrows*) for hydrated $AlPO_4$-11. Peaks associated with aluminum sites experiencing strong interactions with protons from adsorbed water are enhanced by CPDOR. For quadrupolar nuclei subject to sample rotation, the existence of satellite transitions can result in complicated mixings of states, though these can be reduced by use of short CP contact times. Adapted with permission from Ref. [44]

3.4 DOR Averaging of Homonuclear Couplings

In addition to averaging away anisotropic broadening effects from second-order quadrupolar interactions, DOR has recently been shown to reduce line broadening associated with homonuclear dipole–dipole coupling as well [45]. In this case, dipolar line narrowing produced by the rapid rotation of the inner rotor is not destroyed by the relatively slow motion of the large outer rotor, so long as the inner rotor spinning frequency exceeds the strength of the dipole–dipole interactions. The slow spinning outer rotor imposes a slow precessional motion on the spinning axis of the small inner rotor, which can reduce line broadening due to homonuclear dipole–dipole effects. This has been demonstrated by Wu and coworkers [45] for homonuclear ^{23}Na–^{23}Na coupling (≈ 2 kHz) in NaCl, where the cubic symmetry of the sodium sites results in vanishingly small effects from quadrupolar interactions. Averaging of anisotropic homonuclear dipole–dipole interactions in such systems can be achieved even for outer rotor spinning speeds much smaller than the strength of the dipolar interaction. The authors showed furthermore that, under conditions of synchronized DOR, phase reversal of the odd spinning sidebands in this highly degenerate many-body system can be understood without solving the complete many-body problem.

3.5 *Assessment*

From the considerations and examples outlined above, the number and type of molecular conditions amenable to structural investigation by Double Rotation methods are increasing. One constraint on the utility of DOR for structural characterization is the ill-defined upper limit on the magnitude of the quadrupole interaction that is feasible for study. Whereas highly symmetric nuclear environments, experience broadening contributions from higher-order effects that may be small and manageable under high field conditions, such is often not the case for highly asymmetric sites. For example, MAS may be sufficient for obtaining high-resolution NMR spectra for symmetric sites with small quadrupolar coupling constants (small C_Q), while DOR is generally feasible for C_Q up to ca. 10 MHz. Conversely, for highly asymmetric sites (large C_Q), current maximum DOR outer rotor spinning speeds of ca. 1–1.5 kHz, often achievable only under the best of circumstances, are frequently still too slow. In such cases, exceedingly broad spinning sideband manifolds may result, with the desired signal lost in the baseline noise. While synchronized DOR and CPDOR can mitigate this constraint to some extent, broad spinning sideband manifolds tend to be particularly troublesome for quadrupolar nuclei with high atomic mass. The large quadrupolar coupling constants associated with many heavy atomic species often lead to DOR spectra that preclude meaningful analyses. Disordered systems present separate difficulties because distributions of molecular environments, unrelated to intrinsic quadrupolar effects, lead to inhomogeneous broadening that cannot be removed by double rotation. DOR investigations to date have, therefore, focused on applications to lighter NMR-active species, especially ^{17}O, ^{23}Na, and ^{27}Al, in generally ordered local environments. Numerous recent studies have appeared, which have used DOR to probe the molecular structure of an increasingly diverse array of interesting and important inorganic materials containing quadrupolar component species. The expanding list includes ^{11}B DOR of borate glasses [46], ^{17}O DOR of silicate minerals [47], and ^{23}Na DOR of zeolites [48–56, 68], ^{27}Al DOR of porous aluminophosphates [57–67], zeolites [54, 68–70], mesoporous molecular sieves [71], and other minerals or crystalline powders [72–74], and ^{87}Rb DOR of inorganic salts [83].

4 Dynamic-Angle Spinning

As discussed above, Double Rotation eliminates second-order broadening effects in solids by spinning a powdered sample about a time-dependent axis. In this way, the anisotropic terms, which scale as the second- and fourth-Legendre

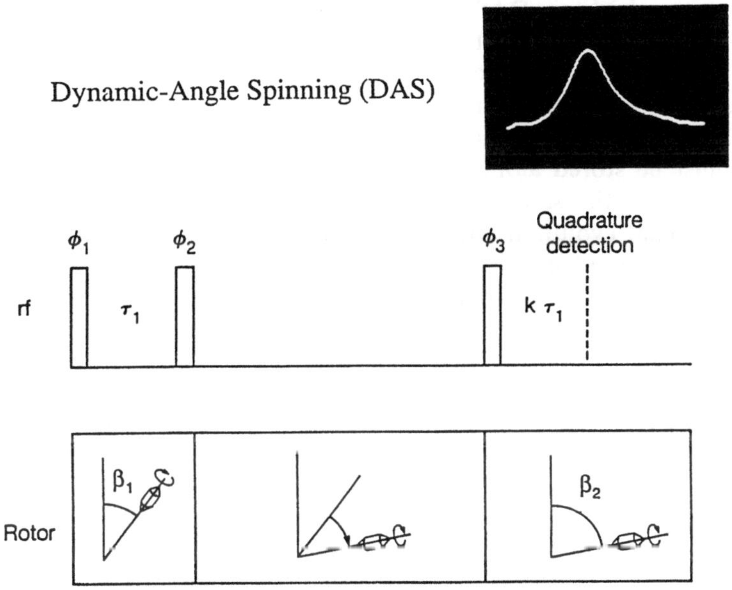

Fig. 10. Schematic diagram of a conventional Dynamic-Angle Spinning experiment showing discrete reorientation of a rapidly spinning sample between two DAS complementary angles β_1 and β_2 synchronized with a series of rf pulses whose phases ϕ_i are described in reference [21]. The DAS experiment refocuses contributions from anisotropic quadrupolar and chemical shift interactions, producing an echo in the time-domain signal whose amplitude is modulated only by isotropic terms. Adapted with permission from Ref. [21]

polynomials $P_2(\cos\beta)$ and $P_4(\cos\beta)$ (Eq. 49) are averaged to zero iteratively: $\langle P_2 \rangle = \langle P_4 \rangle = 0$. The technique of Dynamic-Angle Spinning (DAS), on the other hand, represents an alternate means of eliminating multiple-rank broadening from anisotropic higher-order effects through choice of a different sample trajectory. Whereas in Double Rotation the sample is spun rapidly and continuously about the roots of P_2 and P_4, in the analogous DAS experiment the spinning axis of a rapidly rotating sample is flipped between two discrete angles, which are chosen so that the anisotropic terms in Eq. (49) exactly cancel at the two orientations. Dynamic-Angle Spinning is similar to a conventional MAS experiment, except that the orientation of the spinning axis with respect to the magnetic field $\mathbf{B}_0$ has been made time-dependent in a discrete manner to remove the effects of multiple-rank spectral broadening. This is illustrated schematically in Fig. 10, which shows the simplest case in which the axis of a spinning sample is toggled between two different angles relative to $\mathbf{B}_0$. The choice of angles β_1 and β_2 depends upon the lengths of time τ_1 and τ_2 spent at each orientation, as established by the pair of criteria

$$\tau_1 P_2(\cos\beta_1) = -\tau_2 P_2(\cos\beta_2) \tag{53}$$

and

$$\tau_1 P_4(\cos\beta_1) = -\tau_2 P_4(\cos\beta_2). \tag{54}$$

Figure 10 also shows the basic DAS pulse sequence, which is more complex than that used in DOR. Double Rotation is similar to MAS in that it is a simple pulse-and-acquire experiment. DAS could be as well, except for the finite time needed to switch between the two angles. The mechanical implementation of the switching is discussed below. Here, we note that during the switch, the magnetization must be stored along the z-axis to prevent further evolution; after stable spinning at the second angle is achieved, the magnetization is retrieved for subsequent evolution and detection in the transverse plane. The basic pulse sequence is thus identical to that used in exchange NMR experiments. Furthermore, the effects of the pulse are straightforward to visualize, because the effective second-order spin operators transform like rank-1 and rank-3 tensors, which as noted previously [Sect. 2.2], reduce to just I_z for the central transition.

Unlike in DOR where only angles that are roots of P_2 and P_4 can be used (e.g., 54.74° and 30.56°, respectively), a continuous set of incrementally different angle combinations are possible in DAS, due to the flexibility provided by the time scaling. This is shown in Fig. 11, where the range of DAS complementary angles satisfying the conditions of Eqs. (53) and (54) is plotted as a function of the time spent at the two angles $k_t = \tau_2/\tau_1$. Under these conditions, evolution of the spin system during τ_2, which occurs under the influence of non-zero P_2 and P_4 contributions (Eq. 49), will exactly cancel the evolution that occurs during τ_1. The significance of this is to effect the refocusing of all anisotropic quadrupolar and chemical shift broadening contributions, so that evolution of the magnetization as a function of the total time $t_1 = \tau_1 + \tau_2$ occurs only under the influence of the isotropic terms. This lends itself well to a two-dimensional experiment, whereby the amplitude of the refocused echo in the time domain (Fig. 10) is modulated at isotropic frequencies. Following two-dimensional

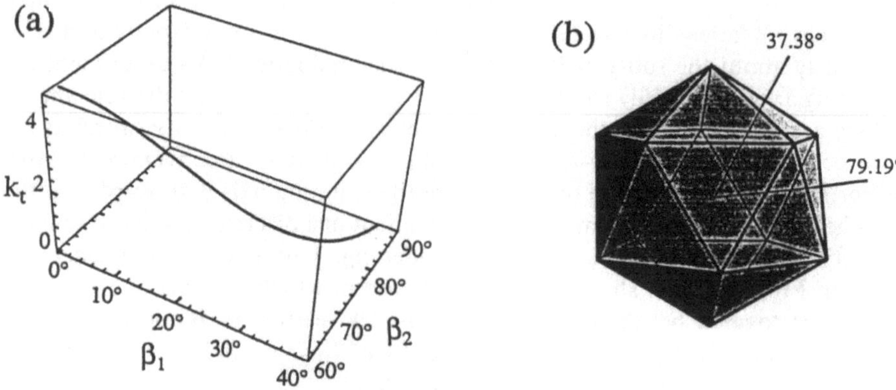

Fig. 11a. Three-dimensional plot of Eqs. (53) and (54) showing the dependence of the DAS complementary angles on the ratio of times $k_t = \tau_2/\tau_1$ spent at the respective orientations β_1 and β_2. **b** When $k_t = 1.0$, equal time is spent at the DAS angles $\beta = 37.38°$ and $\beta = 79.19°$, which corresponds to axes bisecting the faces of an icosahedron. Used with permission from Refs. [5, 20]

Fourier transformation of the time domain data, this leads to narrowed resonance lines in the isotropic F_1 dimension, which can be correlated to anisotropic information remaining in the spectral features of F_2, which in this case reflect the variable-angle spinning line shape at DAS angle β_2.

4.1 DAS Implementation and Probehead Designs

Several different DAS probehead designs have appeared, which strike different balances with respect to various optimization criteria [75]. Desirable, but in some cases conflicting, probehead features include high speed and stability of the discrete sample flip, stable rapid sample rotation, high rf coil filling factor, high and constant resonant circuit efficiencies at different sample orientations, double-resonance capabilities, etc. Figure 12 shows examples of two designs that balance these factors effectively, but in different ways. In the first (Fig. 12a), the stator resides within a fixed coil mounted externally to the entire assembly [76]. The rotor containing the sample spins about an axis normal to that of the stator and rf coil and is supplied with drive and bearing air separately through the stator ends. The axis of the rotor is controlled using a pulley mechanism, which connects the stator assembly to a stepper motor via non-

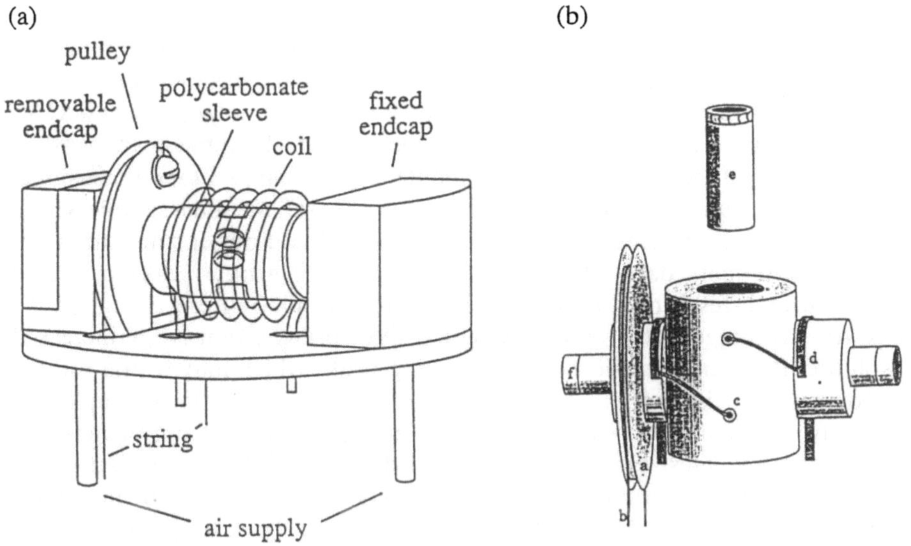

Fig. 12a, b. Diagrams of two probehead designs for the Dynamic-Angle Spinning experiment. **a** The external, fixed solenoid coil arrangement of Mueller et al. permits equivalent excitation and detection efficiency at all angles, though at the expense of a large coil and a reduced filling factor. Used with permission from Ref. [76]. **b** The design of Eastman et al. utilizes a smaller coil wrapped about the stator, which moves with the sample assembly. This configuration has a higher filling factor and is amenable to double-resonance experiments, though probe de-tuning may occur at different DAS angles. Used with permission from Ref. [77]

stretching Kevlar (DuPont) twine. The motor is interfaced with the spectrometer, so that rf pulses are synchronized with the rotor axis flip. Among the advantages of this configuration are tuning stability, which will be the same at all sample orientations, and rapid and stable sample flipping, due to the comparatively low rotational inertia of the stator assembly.

Alternatively, Fig. 12b illustrates a second approach for a DAS probehead designed with different advantages in mind [21, 77]. The principal difference lies in the use of a smaller coil, which is wound within the stator assembly and which therefore undergoes reorientation with the sample. This configuration is based on an earlier design by Terao et al., who developed an angle-switching probehead for spinning samples that was used to separate chemical shift anisotropy patterns and to study heteronuclear dipolar interactions in solids [78, 79]. The design shown in Fig. 12b relies on a pulley connection to the stepper motor, as described above, though other mechanical linkages [80], including a helical screw [38], are also in use. In each case, robust electrical contact to the mobile coil/stator assembly is required. At the expense of diminished tuning stability, which will be different at different sample orientations, the smaller coil provides more versatile rf performance and a significantly larger filling factor.

The first experimental schemes for implementing DAS were conceived by the Saclay and Berkeley groups concurrent with their initial formulations of higher-order averaging theory [18, 19]. In 1989 Chmelka et al. published the first experimental demonstration of DAS, which presented results for ^{17}O in a polycrystalline solid [39]. Following this communication, Mueller et al. published a full description of the method, including one- and two-dimensional applications to ^{23}Na and ^{17}O NMR [21]. Subsequently, a number of important new DAS demonstrations and improvements have appeared, which highlight the versatility and utility of these methods. For example, Fig. 13 shows a two-dimensional ^{17}O DAS spectrum for the mineral wollastonite [47, 81], the same sample whose DOR spectrum is presented in Fig. 7. Two-dimensional Fourier transformation of the time-domain data allows correlation of the isotropic peaks projected along the high-resolution F_1 axis with the anisotropic pattern along the F_2 axis. An advantage of the two-dimensional DAS presentation is the increased resolution that can be achieved by separating peaks which may overlap in the one-dimensional isotropic projection. At 28.2 ppm in Fig. 7, for example, the relatively intense upfield peak arises from overlapping signals from two distinct bridging oxygen sites, which can be discerned in the 2D DAS plot, but which are not resolved in either the accompanying 1D DAS projection or in the 1D DOR spectrum in Fig. 7. The 2D DAS presentation, thus, allows resonances from all nine crystallographically distinct oxygen sites to be resolved for the complicated wollastonite mineral structure. Isotropic ^{17}O DAS line widths of 100–200 Hz are often achievable for polycrystalline powders [47, 81], such as for the wollastonite sample whose spectrum is shown. For moderate quadrupolar coupling constants, the absence of spinning sidebands in the DAS spectrum simplifies the analysis compared to DOR, though at the expense of reduced signal-to-noise which must be tolerated in the 2D DAS measurement.

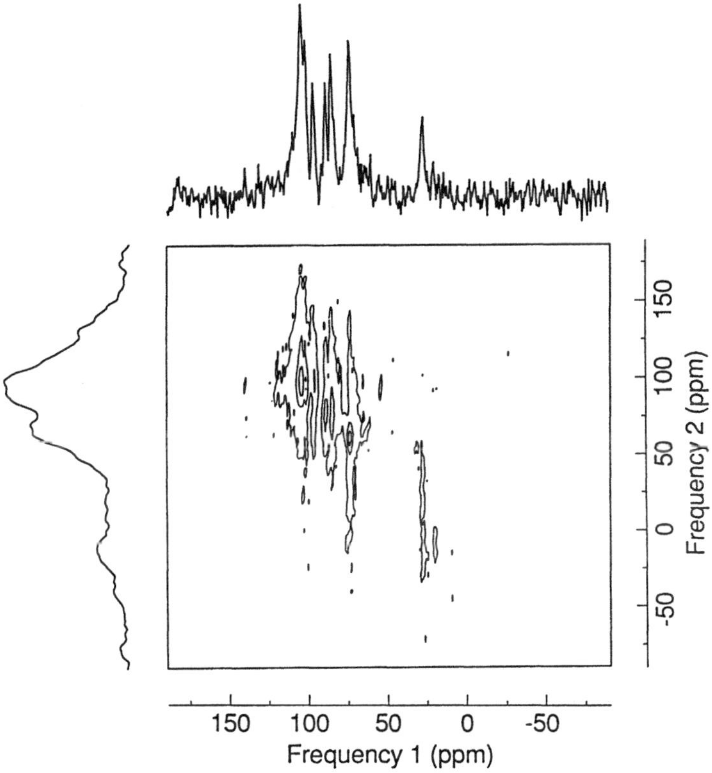

Fig. 13. ^{17}O 2D DAS spectrum of the silicate mineral wollastonite ($CaSiO_3$) at 9.4 Tesla. Isotropic peaks are projected onto the high-resolution F_1 frequency axis, which can be correlated with the anisotropic pattern along F_2 obtained while spinning at angle β_2. The spectra represent absolute-value line shapes acquired using the conventional DAS experiment depicted in Fig. 10. Used with permission from Ref. [47]

4.2 Pure-Absorption-Mode DAS Experiments

Improvements to the original DAS experiment have focused on improving signal sensitivity and resolution, most notably by generating pure-absorption-mode DAS spectra. This is an advancement over the capabilities of the original 2D DAS experiments [21, 47], for which mixed absorption- and dispersion-mode line shapes were obtained that required display of the 2D map as a magnitude spectrum. The consequences of this were altered anisotropic line shapes, diminished resolution, and reduced signal-to-noise, which rendered analysis of the magnitude DAS spectra difficult. Mueller et al. first discussed strategies for overcoming the difficulties associated with such phase-twisted DAS line shapes, whose origins derive from the absence of coherence transfer after formation of the echo in the original DAS experiment [82]. Mueller and coworkers showed how the straightforward addition of a 90° pulse coinciding with the echo maximum at the end of the evolution period effects the transfer of coherence needed for producing pure-absorption-mode DAS line shapes. As a more versatile alternative shown schematically in Figure 14(a), they additionally demonstrated the use of a z-filter storage pulse at the end of the first evolution period, which has the added advantage of permitting the spinner axis to be reoriented during the z-filter to any arbitrary third angle for signal detection. This may be highly desirable, because it permits the DAS signal to be acquired at any convenient angle, unconstrained by previous restrictions to the set of

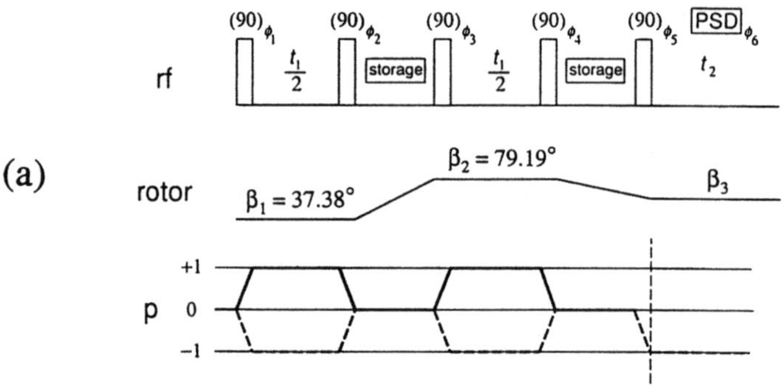

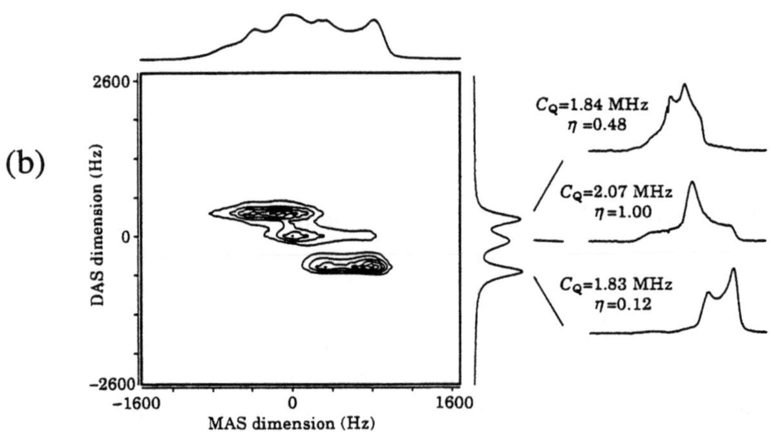

Fig. 14a. Pulse sequence, rotor orientation, and coherence order for a pure-absorption-mode DAS experiment using a z-filter. Used with permission from Ref. [82]. **b** ^{87}Rb DAS spectrum of polycrystalline $RbNO_3$ acquired at 11.7 T using a z-filter DAS sequence with MAS detection. The vertical spectrum shows the 1D DAS projection onto the F_1 axis, which resolves isotropic resonances from three distinct rubidium sites. F_2 slices through each of the isotropic peaks are MAS $P_4(\cos\beta)$ powder patterns, which can be combined with simulations to establish quadrupolar coupling parameters and isotropic chemical shift values for each site. Adapted with permission from Ref. [83]

DAS complementary angles shown in Fig. 11. Detection at the magic angle, for example, is impossible in a conventional DAS experiment, because 54.74° is not an allowed DAS angle. The use of the DAS z-filter allows isotropic resonances of quadrupolar species to be measured, while still permitting their correlation with the respective anisotropic line shapes at the magic angle. This may be particularly convenient, since quadrupolar broadening in the anisotropic F_2 dimension can then be probed in isolation from other anisotropic first-order contributions. For example, while chemical shift anisotropy effects may complicate DAS analyses at detection angles other than 54.74°, they will be averaged to zero by rapid rotation during DAS detection at the magic angle.

Baltisberger et al. [83] have illustrated the efficacy of the z-filter DAS approach for a series of polycrystalline rubidium salts, for which ^{87}Rb pure-absorption-mode DAS spectra were acquired with signal detection at the magic angle. As shown in Fig. 14b, three isotropic peaks are resolved along the vertical F_1 axis, each corresponding to a distinct rubidium site. Taking slices parallel to the anisotropic F_2 axis through each of the isotropic peaks yields MAS powder patterns, which permit the quadrupolar coupling parameters and isotropic chemical shift for each site to be established independently. Moreover, in this study, the authors performed DAS experiments at several magnetic field strengths, which substantially increases the accuracy of the measurements [83]. The capability of DAS to correlate isotropic and anisotropic spectral features is a distinct advantage over Double Rotation averaging methods discussed earlier, which can provide only isotropic frequency information.

Grandinetti et al. [84] have recently described several DAS experiments that demonstrate alternate means of obtaining pure-absorption-mode 2D DAS spectra with the concomitant benefit of enhanced sensitivity over that provided by other DAS techniques. For example, though the pure-absorption-mode DAS spectra arising from use of a z-filter show improved resolution, the extra storage pulse needed for the z-filter offsets the $\sqrt{2}$ enhancement that would normally accrue for a pure-phase (vis-à-vis a mixed-phase) spectrum.

Grandinetti and coworkers modified the DAS experiment by redefining the evolution times, so that acquisition of the signal begins immediately after the final pulse, as opposed to the top of the echo [84]. The resulting spectrum is related to the conventional DAS plot by a shearing transformation (in the form of a t_1-dependent, first-order phase correction), which correlates the 1D Variable-Angle Spinning (VAS) spectra associated with each rotor angle in the DAS experiment. This is related to 2D exchange techniques [4, 85] with rotor reorientation during the mixing time and leads to an amplitude-modulated response in the t_1 domain. Employing hypercomplex 2D data acquisition [4, 86] with phase separation during t_1 results in pure-absorption-mode 2D DAS spectra with improved sensitivity over the original DAS experiments. For disordered systems in which there is a continuous distribution of atomic environments (see Sect. 4.3 below), Grandinetti et al. [84] further noted that combining t_1 phase separation with a time-shifted DAS echo in t_2 produces still greater enhancements in signal-to-noise, up to a factor of $2\sqrt{2}$. The pulse sequence

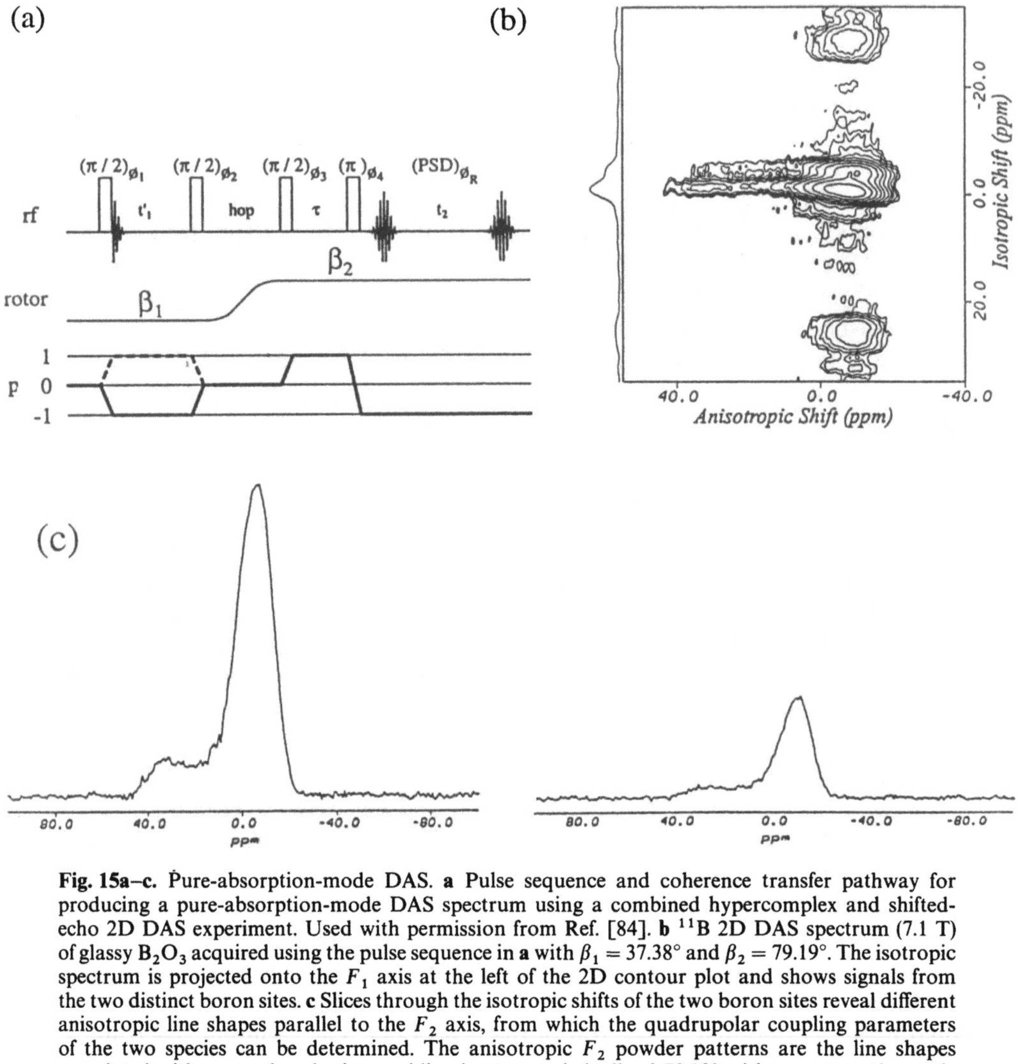

Fig. 15a–c. Pure-absorption-mode DAS. **a** Pulse sequence and coherence transfer pathway for producing a pure-absorption-mode DAS spectrum using a combined hypercomplex and shifted-echo 2D DAS experiment. Used with permission from Ref. [84]. **b** ^{11}B 2D DAS spectrum (7.1 T) of glassy B_2O_3 acquired using the pulse sequence in **a** with $\beta_1 = 37.38°$ and $\beta_2 = 79.19°$. The isotropic spectrum is projected onto the F_1 axis at the left of the 2D contour plot and shows signals from the two distinct boron sites. **c** Slices through the isotropic shifts of the two boron sites reveal different anisotropic line shapes parallel to the F_2 axis, from which the quadrupolar coupling parameters of the two species can be determined. The anisotropic F_2 powder patterns are the line shapes associated with a sample spinning rapidly about an axis inclined 79.19° with respect to the static $\mathbf{B}_0$ field. Used with permission from Ref. [87]

and coherence transfer pathway for this experiment are shown in Fig. 15a. This DAS experiment has been applied by Youngman and Zwanziger to the structural characterization of borate glass, where inhomogeneous spectral broadening has previously been an obstacle to meaningful analysis [87]. Figure 15b shows a ^{11}B pure-absorption-mode 2D DAS spectrum of glassy B_2O_3 in which two distinct boron sites, corresponding to boroxol rings and trioxide groups, are resolved in the projection onto the isotropic F_1 frequency axis. One-dimensional slices through each of the isotropic shifts (Fig. 15c) display distinctly different anisotropic powder patterns from which different quadrupolar and chemical shift parameters are established. These experiments illustrate the utility of the hypercomplex, full-echo-acquisition variant of DAS for obtaining pure-absorption-mode 2D DAS spectra with high resolution and enhanced sensitivity.

4.3 Isotropic/Anisotropic DAS Correlations in Disordered Materials

While DAS (and DOR) provide the narrowest lines for highly crystalline samples, recent experiments reflect the utility of DAS experiments for quantifying structural features of less ordered materials, particularly inorganic glasses. In the case of a sample possessing an appreciable degree of disorder at a molecular level, broadening of DAS (and DOR) line shapes will result from a dispersion of shifts, rather than intrinsic anisotropic nuclear spin interactions. Though

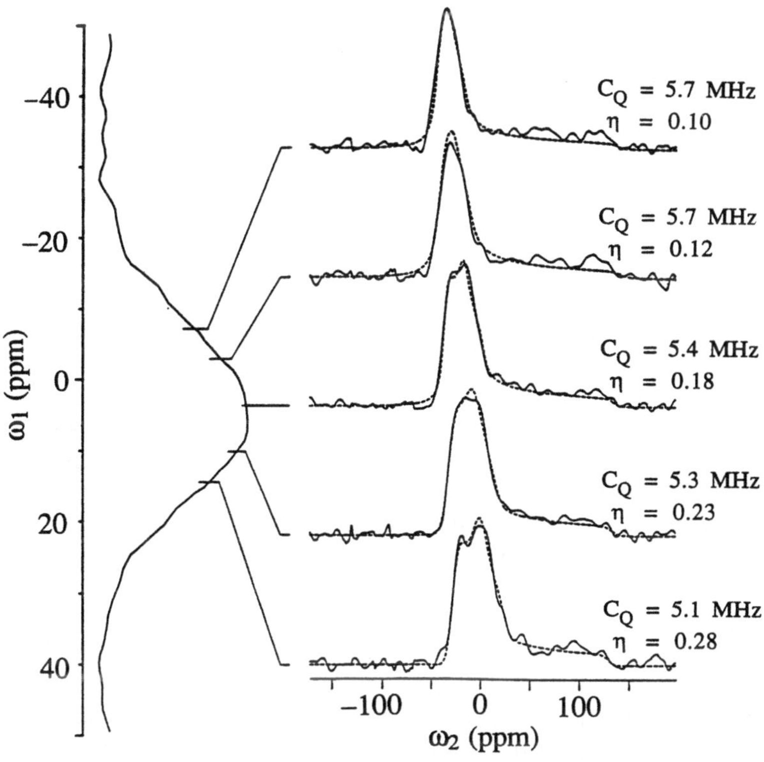

Fig. 16. ^{17}O 1D DAS spectrum (9.4 T) showing, along the vertical, the broadened isotropic resonance associated with bridging oxygen atoms in potassium silicate ($K_2Si_4O_9$) glass enriched to 43% in ^{17}O. Slices parallel to the anisotropic F_2 dimension are shown, accompanied by dashed fits, which provide the magnitude and asymmetry of the electric field gradient at different points across the isotropic line shape. Used with permission from Ref. [88]

DAS line widths may be an order of magnitude broader than those obtained for crystalline materials, they can result solely from distributions of molecular environments and thus reflect the degree of disorder associated with different sites in a material. This has been shown by Farnan et al. [88] for distributions of oxygen sites in potassium silicate glasses using ^{17}O DAS and more recently by Youngman and Zwanziger [87], as discussed above (Fig. 15b), for distributions of boron sites in borate glasses using ^{11}B DAS. As in the case of the polycrystalline sample in Fig. 14b, anisotropic line shapes in the glass studies were analyzed for different F_2 slices taken at a collection of frequency positions across the isotropic F_1 axis. For disordered materials, the normally high-resolution F_1 dimension may be inhomogeneously broadened by a distribution of molecular environments, as shown in the ^{17}O DAS spectrum in Fig. 16 for ^{17}O-enriched potassium silicate ($K_2Si_4O_9$) glass. By fitting different anisotropic F_2 spectral slices across F_1 (see dashed lines in Fig. 16), Farnan et al. measured the variation of the asymmetry parameter η among locally different bridging oxygen species that were distinguished by their F_1-selected isotropic shifts [88]. The asymmetry parameter η describes the ^{17}O electric field gradient (EFG) at a given oxygen site, the magnitude of which is correlated closely with the Si—O—Si bond angle [89, 90]. Thus, the variation of η across the broadened isotropic peak reflects the distribution of Si—O—Si bond angles in the disordered sample.

4.4 Sideband Manipulations in DAS

The spinning sidebands that appear in DAS arise from the periodic time dependence introduced to the Hamiltonian by spinning the sample, which must be rotating faster than the homogeneous width of the line. While sidebands usually appear at integer multiples of the spinning frequency and with diminishing intensity farther from the center band, certain types of two-dimensional NMR experiments exhibit more complex sideband behavior. This is especially true for 2D echo techniques, including 2D J-Spectroscopy and Dynamic-Angle Spinning, which rely on an evolution period that is divided into parts of generally unequal length. In 2D J-Spectroscopy, sidebands appear at multiples of one-half the spinning frequency in the F_1 dimension, because the two periods of evolution are of equal duration, separated by a π pulse. In DAS, however, the relative lengths of the two evolution period components depend on the choice of DAS angles, as accounted for by the time ratio k_t needed to refocus first- and second-order anisotropic contributions to the time-domain echo (Fig. 11a). Consequently, the refocusing period $x_2 t_1$ shown in Fig. 17a will not in general be equal to the dephasing time $x_1 t_1$, with the result that complicated spinning sideband patterns may be expected. For example, sideband spacing at noninteger multiples of the rotor spinning speed [91], together with complex intensity variations [20], can be observed in the isotropic dimension of DAS spectra. Such observations have been discussed in detail by Sun and coworkers [20] and Grandinetti et al. [91], who have described the origin and manipulation of DAS spinning sideband manifolds.

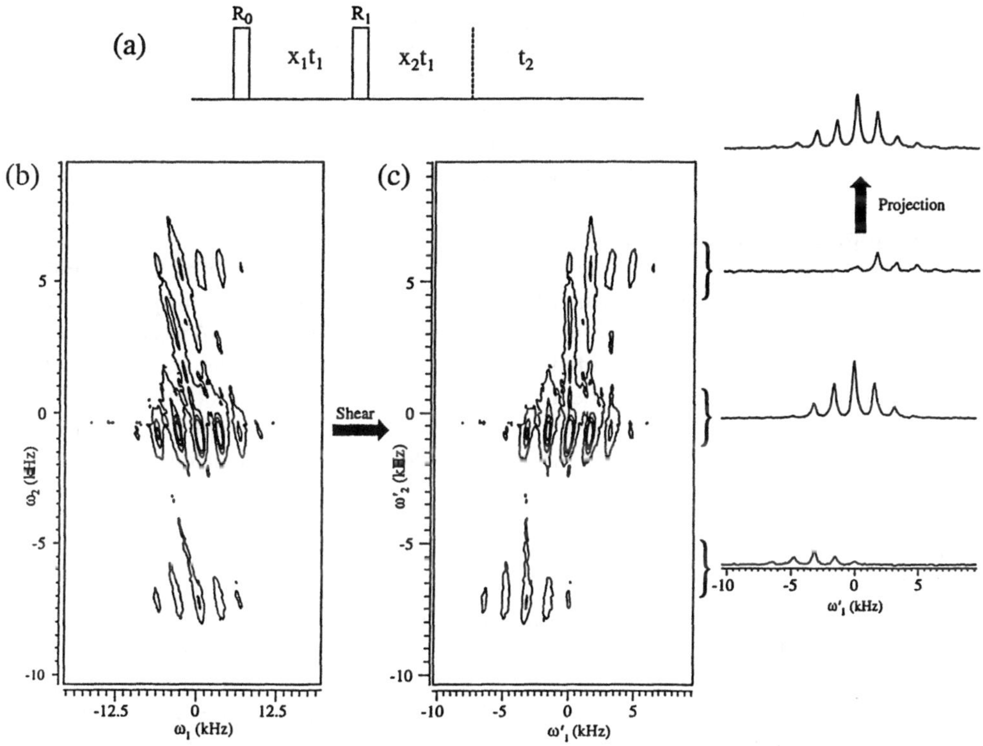

Fig. 17a. DAS pulse sequence showing the two not necessarily equal periods $x_1 t_1$ and $x_2 t_1$ that together comprise the evolution time. **b** ^{87}Rb 2D VAS spectrum of $RbClO_4$ at 9.4 Tesla correlating the anisotropic spectra of the DAS complementary angles $\beta_1 = 37.38°$ and $\beta_2 = 79.19°$ ($k_t = 1$). **c** The 2D VAS-correlated spectrum is converted into the 2D DAS spectrum using a shear transformation. For this case, the times spent at the two angles are the same ($x_1 = 0.5$ and $x_2 = 0.5$), so that the spinning sidebands are aligned with respect to the isotropic F_1 axis and occur at $0.5\,\omega_r = 1.6$ kHz. For other DAS angles, the sidebands will not align with respect to F_1, resulting in a more complicated manifold. Used with permission from Ref. [91]

These works make clear the connection between DAS and correlated Variable-Angle Spinning (VAS) experiments [91]. By correlated VAS we mean experiments that involve a change of the rotor axis between two (or more) angles during the mixing time of a multidimensional NMR experiment; DAS is the special case where the angles are defined by Eqs. (53) and (54). [Correlated VAS should not be confused with Variable-Angle Correlation Spectroscopy (VACSY), a different type of isotropic-anisotropic experiment developed by Frydman et al. [92, 93] (see also [94]), which is thus far applicable only to spin-$\frac{1}{2}$ nuclei.] The connection between correlated VAS and DAS arises through a shearing transformation, as mentioned in Sect. 4.2 above. For example, Fig. 17b contains a ^{87}Rb 2D NMR spectrum of rubidium chlorate ($RbClO_4$), which correlates VAS spectra acquired at the DAS angles 37.38° and 79.19° ($k_t = 1$) and in which the spinning sidebands appear at multiples of the rotor frequency [91]. The correlated 2D VAS plot is converted into the 2D DAS spectrum in Fig. 17c by applying a 45° shearing transformation and by scaling the resultant plot by $x_1 = x_2 = 0.5$ (for $k_t = 1$). This leads to a DAS spectrum in which all spinning sidebands are aligned in the F_1 dimension, with the result that they appear at equally spaced intervals of one-half the rotor speed ω_r in the isotropic projection. For other DAS angles, however, the shearing and scaling operations produce sideband patterns that may not align, so that in general complicated and often overlapping manifolds result.

One means of circumventing the difficulties presented by complex sideband patterns is to design an experiment in which the sidebands are suppressed or eliminated entirely. As mentioned earlier, MAS, DOR, and DAS sample reorientation methods are founded on the need to average the second- and fourth-rank tensors that describe anisotropic nuclear spin interactions in solids. The spatial dependences of these tensors, as we have said, display cubic and icosahedral symmetries (Fig. 5), which provide intuition and insight on these and other possible line narrowing trajectories. This can be seen, for example, from the equivalent perspective of a sample-fixed frame of reference in which the field $\mathbf{B}_0(t)$ reorients during the experiment (rather than the sample). Figure 18a, for example, portrays the cubic symmetry and $\mathbf{B}_0(t)$ trajectory associated with Magic-Angle Spinning/Hopping experiments (Fig. 1) from a frame of reference fixed with respect to the sample. In this representation, the time-dependent field $\mathbf{B}_0(t)$ sweeps through the vertices of the octahedron in a conical path, fulfilling the necessary $P_2(\cos \beta) = 0$ averaging condition. Similarly, a DAS trajectory in which equal time is spent at the angles 37.38° and 79.19° ($k_t = 1$) can be considered a path in the sample-fixed frame traversed by the magnetization on two cones, whose bases connect the vertices of a dodecahedron (Fig. 18b), and whose apices reside at the dodecahedron's center. A second important pair of DAS complementary angles are 0.00° and 63.43°, which require the largest time-scaling factor $k_t = 5$, corresponding to a trajectory that can be described by an icosahedron, as shown in Fig. 18c. All other DAS solutions likewise relate to the symmetries of a dodecahedron and/or icosahedron [20, 20a, 32].

In analogy to Magic-Angle Hopping, Gann et al. have demonstrated a scheme for Dynamic-Angle Hopping (DAH), whereby sideband-free DAS spectra can be obtained [95]. A key objective of their investigation centered on preservation of signal intensity: storage/restoration pulses that could be employed prior to each sample hop store only half of the transverse magnetization, so that under these conditions, signal-to-noise will be reduced by a factor of $\sqrt{2}$ for *each* discrete sample reorientation. The need for multiple sample hops is, thus, accompanied by the prospect of severe sensitivity difficulties unless the losses of signal imposed by the storage pulses can be mitigated. In a Dynamic-Angle Hopping experiment, a minimum of six discrete sample orientations must be visited to achieve the quadrupolar line narrowing desired, in this case, five 72° steps about an axis oriented at 63.43° with respect to $\mathbf{B}_0$ followed by repositioning of the sample rotation axis to an orientation parallel to $\mathbf{B}_0$ (0°). As shown in Fig. 18c, this trajectory samples the vertices of an icosahedron, as

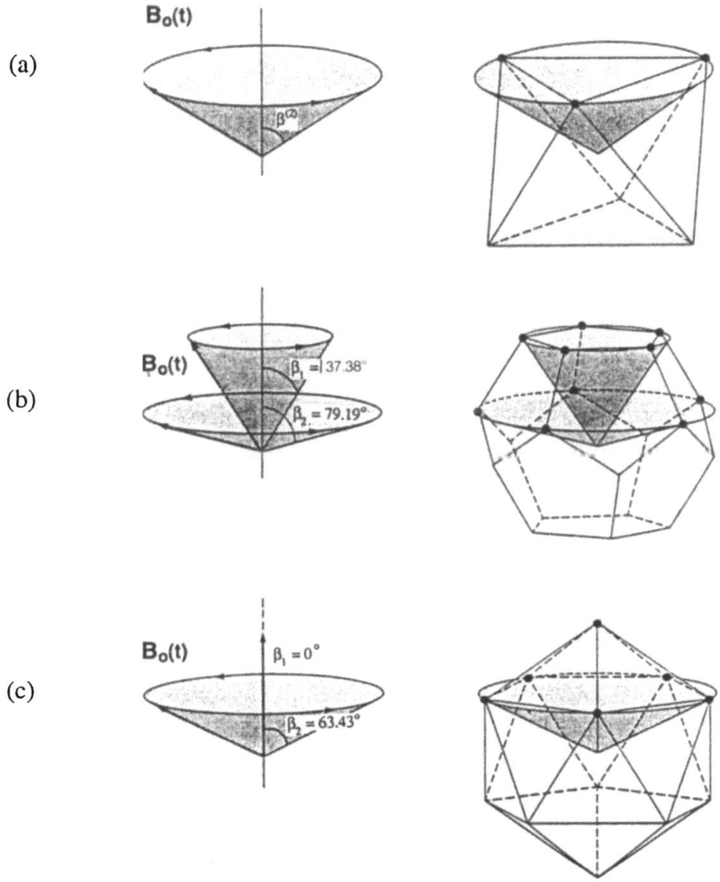

Fig. 18a–c. Symmetries of various NMR averaging techniques viewed in sample-fixed frames of reference in which the time-dependent magnetic field $\mathbf{B}_0$ is reoriented through the conical trajectories shown. **a** In Magic-Angle Hopping, the data are acquired discretely at three field orientations directed toward the three vertices of the face of an octahedron, while Magic-Angle Spinning corresponds to continuous rapid reorientation through the same points. In DAS and Dynamic-Angle Hopping, it is necessary to cross: **b** ten vertices of a dodecadedron, when complementary angles $\beta_1 = 37.38°$ and $\beta_2 = 79.19°$ are used and the times spent on the two cones are the same ($k_t = 1$); or **c** six vertices of an icosahedron for complementary angles $\beta_1 = 0.00°$ and $\beta_2 = 63.43°$ with $k_t = 5$. Adapted with permission from Ref. [20]

demanded by the icosahedral symmetry on which these experiments are founded (Fig. 5 and Fig. 11b). Gann and coworkers [95] circumvented the threat of diminished DAH sensitivity by applying a series of DAS-synchronized π pulses to a rapidly spinning sample, thereby allowing all components of the evolving magnetization to be retained. Termed "DAH-180", this approach effectively mimics the *discrete* (63.43°, 0°) path described above, while preserving signal-to-noise through avoidance of culpable storage pulses. Figure 19(a) demonstrates the effectiveness of the DAH-180 experiment, showing a high-resolution side-

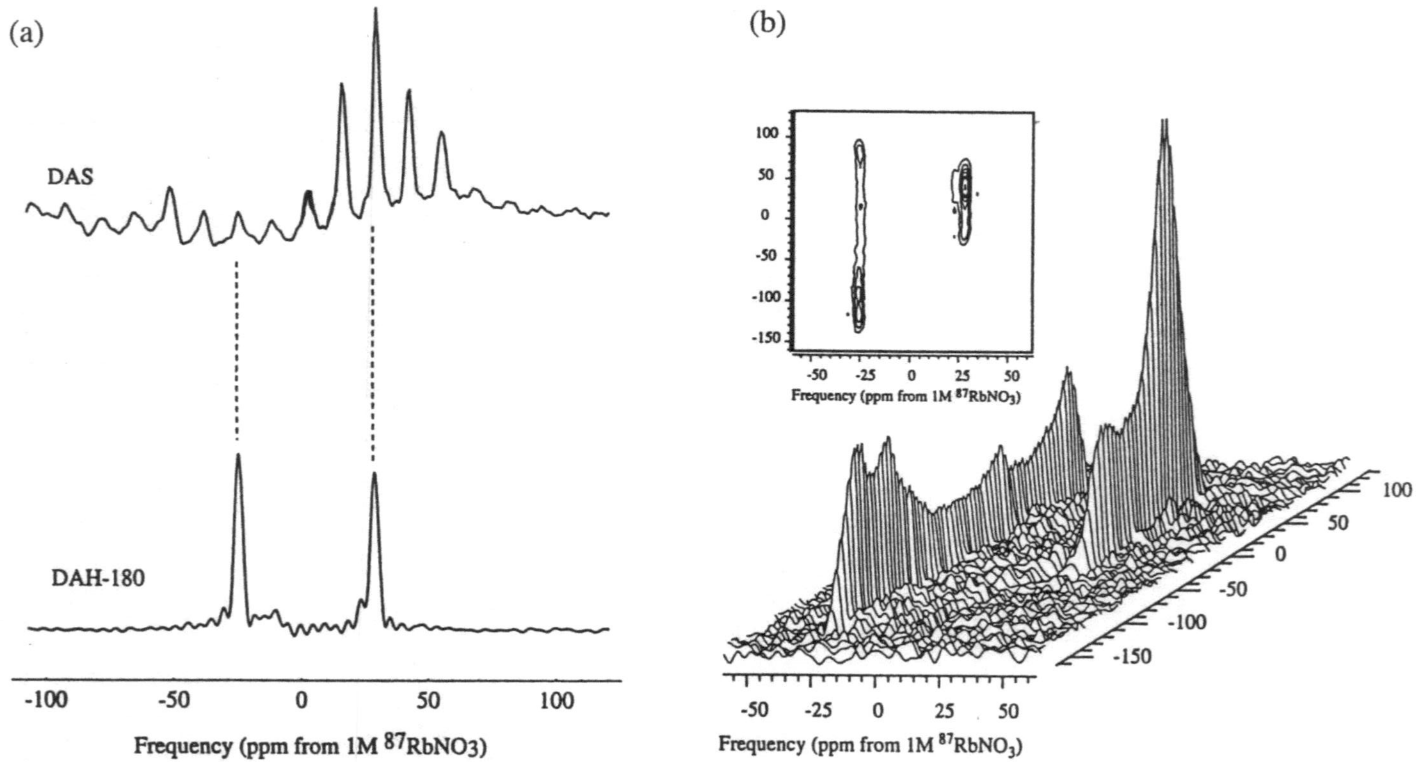

Fig. 19a. ^{87}Rb 1D DAS and 1D Dynamic-Angle Hopping DAH-180 spectra of Rb_2SO_4 at 9.4 Tesla. The sideband-free DAH-180 spectrum was obtained using rotor-synchronized π pulses to preserve signal intensity. **b** 2D stacked and contour plots of the DAH-180 spectrum in **a** of Rb_2SO_4 showing the narrow isotropic line widths in the F_1 dimension and the full static powder patterns for each resolved site in the anisotropic F_2 dimension. Used with permission from Ref. [95]

band-free isotropic spectrum of rubidium sulfate (Rb_2SO_4) in which full signal intensities of the two ^{87}Rb sites are completely preserved. As in the DAS analyses described previously, the ^{87}Rb 1D DAH-180 spectrum is the projection onto the isotropic F_1 axis. Unlike DAS, however, the anisotropic F_2 pattern (Fig. 19b) contains the full static powder line shapes, reminiscent of the second dimension in Magic-Angle Hopping spectra [16, 17]. As is clear from Fig. 19a, DAH-180 leads to a significantly simpler isotropic spectrum than that provided by DAS for the same Rb_2SO_4 sample. The result is a versatile correlation between isotropic and anisotropic line shapes, from which previously difficult-to-extract structural information can be obtained through judicious adaptation of the flexible DAS technique.

4.5 Double Resonance DAS

As with DOR, the range of applications of DAS can be extended by coupling double resonance strategies with spatial modulation. Examples of DAS, in conjunction with the standard double-resonance experiments of high-power decoupling and cross-polarization, have been demonstrated using double-tuned DAS probeheads. As discussed in Sect. 4.1, such probehead designs must balance often conflicting criteria, which can influence dramatically the double-resonance performance of a device. For example, the design of Eastman and coworkers shown in Fig. 12b relies on a small rf coil that moves with the sample as its rotation axis is flipped [77]. This arrangement provides a good filling factor, but requires care to ensure stable tuning as the sample/coil orientation is changed. Such difficulties are tractable, however, as demonstrated by Eastman et al., who have used this design to obtain ^{1}H-decoupled DAS spectra of ^{11}B in boric acid [77].

Alternatively, the probehead design of Mueller et al. shown in Fig. 12a, relies on a large stationary coil, which surrounds the stator as well as the spinning sample. This configuration possesses a lower filling factor, but provides stable tuning, since the coil remains fixed during the sample flip [76]. With such a design, it is also possible to apply rf pulses to a sample while it rotates rapidly about an axis oriented along the static magnetic field, i.e. at $\beta = 0°$ with respect to $\mathbf{B}_0$. This is an important consideration for achieving cross-polarization in conjunction with DAS (CPDAS), an experiment for which the spin dynamics can be quite complex. Spin-locking a quadrupolar nucleus while rapidly rotating the sample about a single axis, followed by cross-polarization from spin-$\frac{1}{2}$ nuclei to such a nucleus, pose formidable problems that have been treated recently by Vega [96, 97]. The advantage of spinning the sample at $\beta = 0°$ is that this is equivalent to not spinning the sample at all, with the result that the complexity of the spin dynamics can be significantly reduced. Moreover, as a solution to Eqs. (53) and (54), 0° is a DAS complementary angle linked to 63.43° by the proportionality factor $k_t = 5$. Baltisberger et al. [98] have exploited this strategy in experiments on proton cross-polarization to ^{23}Na, showing ^{23}Na DAS spectra with significantly enhanced signal. These authors also demonstrated that cross-polarization at $\beta = 0°$, followed by data acquisition at any arbitrary spinning angle, yields superior line shape fidelity over normal CPMAS or CPVAS for quadrupolar nuclei. The same approach has also been used by Gann et al. [99] to enhance ^{17}O signal intensity in the DAS spectrum of ^{17}O-enriched alanine by cross-polarization from protons.

In a related series of experiments, Fyfe and coworkers [100] have reversed the direction of polarization transfer, using a (0°, 54.74°) sample trajectory to enhance sensitivity from spin-$\frac{1}{2}$ species (^{31}P) by efficient cross-polarization from quadrupolar nuclei (^{27}Al). As for the CPDAS examples discussed above, the axis of a rapidly rotating sample was inclined initially at an angle $\beta = 0°$ with respect to the $\mathbf{B}_0$ field to facilitate contact among the heteronuclear spin species. After flipping the rotation axis of the sample to the magic angle, the FID was acquired and the data subsequently Fourier-transformed to obtain a high-resolution spectrum of the spin-$\frac{1}{2}$ ^{31}P species. This experiment has the advantage that the repetition rate for the sequence is set by spin-lattice relaxation of the quadrupolar nuclei, rather than the (typically long) T_1 of the spin-$\frac{1}{2}$ species. While the objective of this experiment is signal enhancement from spin-$\frac{1}{2}$ species, as opposed to averaging higher-order interactions, it provides a good illustration of the power and flexibility of this class of sample reorientation schemes.

5 Concluding Remarks

In comparing the experimental options available for narrowing quadrupolar lines in solids, Variable-Angle Spinning, Double Rotation, and Dynamic-Angle Spinning have different advantages and limitations based on their respective effectiveness, versatility, and convenience of implementation. While rotating a sample at high speed about a single axis is routine and easily achieved, it is not possible to average completely anisotropic broadening contributions to 1D VAS spectra. Both Double Rotation and Dynamic-Angle Spinning, on the other hand, are capable of averaging first- and second-order anisotropic interactions to zero, so that isotropic lines are obtained in DOR and DAS spectra, often with high resolution.

Of these methods, Double Rotation is the most difficult to implement mechanically, requiring delicate engineering and careful control of the coupled rotor system for stable operation. Several technical limitations currently constrain DOR to investigations at room temperature and on abundant species, primarily because of thermal-induced changes in the delicately machined rotor components and poor sample filling factors, respectively. Current designs permit outer rotor spinning frequencies of about 1 kHz, which in practice limits the utility of DOR to species with moderate quadrupolar coupling constants below ca. 10 MHz, for which manageable spinning sideband patterns are produced. Because DOR is inherently a 1D experiment, it provides no T_1 relaxation restrictions and, depending on the degree of signal averaging required, can yield acceptable signal-to-noise over periods of minutes or hours.

Dynamic-Angle Spinning is mechanically straightforward to perform and provides impressive versatility in terms of sample trajectories and pulse cycles. Its inherently 2D format requires substantially longer measurement times compared to DOR, but provides highly useful information that correlates isotropic frequency shifts with corresponding anisotropic powder line shapes. The versatility and correlation capability of DAS are its most important advantages. It is constrained by currently ca. 30-ms flip times, which require longer relaxation times for feasible DAS study. This restriction is especially troublesome for ^{27}Al species, a number of which have been examined in detail using DOR (see Sect. 3.5), but which have short T_1's that have generally proven infeasible for investigation by DAS (an exception is for the mineral petalite discussed in reference [75]). Compared to DOR, fast single-axis spinning in DAS produces fewer sidebands, which can be diminished further by using rotor synchronization. Future developments are expected to lead to new probeheads designed for more complicated trajectories, as would be required to average interactions beyond second-order. The connection between correlated variable-angle spectra and DAS promises to provide a synergistic foundation for new developments and continuing improvements in these flexible techniques.

Manifested by the invention of ingenious new methods for the study of quadrupolar nuclei, the past several years have witnessed tremendous progress in overcoming technical difficulties associated with structural characterization of quadrupolar species by solid-state NMR. This has led to and is being encouraged by resultant new discoveries in areas of materials chemistry and structure. At the moment, higher-order averaging methods, such as Double Rotation and Dynamic-Angle Spinning, are continuing to experience rapid and simultaneous development, demonstration, improvement, and widening application utility. We anticipate that this rapid advancement will be sustained and broadened, especially in burgeoning areas of new materials applications, as the accessibility of these sophisticated methods continues to increase.

Acknowledgments. The authors are indebted to their colleagues and coworkers, particularly those at Berkeley, for much of the work that is contained herein. BFC acknowledges funding support from the Camille and Henry Dreyfus Foundation, the David and Lucile Packard Foundation, and the U.S. National Science Foundation N.Y.I. program under Grant No. DMR-9257064. JWZ acknowledges support from the U.S. National Science Foundation under Grant No. DMR-9115787.

6 References

1. Pines A, Gibby MG, Waugh JS (1973) J Chem Phys 59: 569
2. Haeberlen U (1976) Adv Magn Reson Suppl. 1, Waugh JS (ed) Academic, New York
3. Mehring M (1983) Principles of high resolution NMR in solids, 2nd edn. Springer, Berlin Heidelberg New York
4. Ernst RR, Bodenhausen G, Wokaun A (1987) Principles of nuclear magnetic resonance in one and two dimensions. Clarendon, Oxford
5. Wooten EW, Mueller KT, Pines A (1992) Accts Chem Res 25: 209
6. Virlet J (1992) J Chim Phys 89: 359
7. Freude D, Haase J (1993) In: NMR basic principles and progress, Vol. 29. Springer, Berlin Heidelberg New York, p 1
8. Jäger C (1994) In: NMR basic principles and progress, Vol. 31. Springer, Berlin Heidelberg New York, p 133
9. Andrew ER, Bradbury A, Eades RG (1958) Nature 182: 1659
10. Lowe IJ (1959) Phys Rev Lett 2: 285
11. Joshi AW (1984) Matrices and tensors in physics, 2nd edn. Wiley, New York
12. Maricq MM, Waugh JS (1979) J Chem Phys 70: 3300
13. Goldman M, Grandinetti PJ, Llor A, Olejniczak Z, Sachleben JR, Zwanziger JW (1992) J Chem Phys 97: 8947
14. Edmonds AR (1960) Angular momentum in quantum mechanics. Princeton University, Princeton, New Jersey
15. Varshalovich DA, Moskalev AN, Khersonskii VK (1988) Quantum theory of angular momentum. World Scientific, Singapore
16. Bax A, Szeverenyi NM, Maciel GE (1983) J Magn Reson 52: 147
17. Szeverenyi NM, Bax A, Maciel GE (1985) J Magn Reson 61: 440
18. Llor A, Virlet J (1988) Chem Phys Lett 152: 248
19. Virlet J, Ninth European Experimental NMR Conference, Bad-Aussee, Austria, 1988; Pines A, Ninth European Experimental NMR Conference, Bad-Aussee, Austria, 1988; Chingas GC, Lee CJ, Lippmaa E, Mueller KT, Pines A, Samoson A, Sun BQ, Suter D, Terao T (1988) In: Stankowski J, Piślewski N, and Idziak S (eds) Proceedings, XXIV Congress Ampère, Poznań. D62
20. Sun BQ, Baltisberger JH, Wu Y, Samoson A, Pines A (1992) Solid State Nucl Magn Reson 1: 267

20a. Emsley L, Pines A (1993) In: Maraviglia B (ed) Nuclear magnetic resonance, Proceedings of the International School of Physics (Enrico Fermi). North-Holland, Amsterdam, p 123

21. Mueller KT, Sun BQ, Chingas GC, Zwanziger JW, Terao T, Pines A (1990) J Magn Reson 86: 470
22. Samoson A, Lippmaa E (1983) Chem Phys Lett 100: 205
23. Samoson A (1985) Chem Phys Lett 119: 29
24. Skibsted J, Nielsen NC, Bildsøe H, Jakobsen HJ (1991) J Magn Reson 95: 88; Skibsted J, Nielsen NC, Bildsøe H, Jakobsen HJ (1992) Chem Phys Lett 188: 405
25. Jäger C (1992) J Magn Reson 99: 353; Kunath G, Losso P, Steuernagel S, Schneider H, Jäger C (1992) Solid State Nucl Magn Reson 1: 261; Jäger C, Kunath G, Losso P, Scheler G (1993) Solid State Nucl Magn Reson 2: 73
26. Samoson A, Kundla E, Lippmaa E (1982) J Magn Reson 49: 350
27. Ganapathy S, Schramm S, Oldfield E (1982) J Chem Phys 77: 4360
28. Lefebvre F, Amoureux JP, Fernandez C, Derouane EG (1987) J Chem Phys 86: 6070
29. Dec SF, Maciel GE (1990) J Magn Reson 87: 153; Alemany LB (1993) Appl Magn Reson 4: 179
30. Skibsted J, Bildsøe H, Jakobsen HJ (1991) J Magn Reson 92: 669
31. Zheng Z, Gan Z, Sethi NK, Alderman DW, Grant DM (1991) J Magn Reson 95: 509
32. Samoson A, Sun B, Pines A (1992) In: Bagguley DMS (ed) Pulsed magnetic resonance: NMR, ESR, and optics. Oxford
33. Samoson A, Lippmaa E, Pines A (1988) Molec Phys 65: 1013
34. Samoson A, Pines A (1989) Rev Sci Instrum 60: 3239
35. Wu Y, Sun BQ, Pines A (1990) J Magn Reson 89: 297
36. Bruker Analytische Messtechnik, D-76287 Rheinstetten, Germany
37. Chemagnetics/Otsuka Electronics, Fort Collins, Colorado, 80525 USA

38. Doty Scientific, Inc., Columbia, South Carolina, 29223 USA
39. Chmelka BF, Mueller KT, Pines A, Stebbins J, Wu Y, Zwanziger JW (1989) Nature 339: 42
40. Samoson A, Lippmaa E (1989) J Magn Reson 84: 410
41. Samoson A (1993) Chem Phys Lett 214: 456
42. Cochon E, Amoureux JP (1993) Solid State Nucl Magn Reson 2: 205
43. Amoureux JP, Cochon E (1993) Solid State Nucl Magn Reson 2: 223
44. Wu Y, Lewis D, Frye JS, Palmer AR, Wind RA (1992) J Magn Reson 100: 425
45. Wu Y, Peng ZY, Olejniczak Z, Sun BQ, Pines A (1993) J Magn Reson, Ser A 102: 29
46. Youngman RE, Zwanziger JW, Janicke M, Chmelka BF (manuscript in preparation)
47. Mueller KT, Wu Y, Chmelka BF, Stebbins J, Pines A (1991) J Am Chem Soc 113: 32
48. Jelinek R, Pines A, Özkar S, Ozin GA (1991) Nanotechnology 2: 182
49. Jelinek R, Özkar S, Ozin GA (1992) J Phys Chem 96: 5949; Jelinek R, Özkar S, Ozin GA, Pastore HO (1992) J Phys Chem 96: 9582
50. Jelinek R, Chmelka BF, Stein A, Ozin GA (1992) J Phys Chem 96: 6744
51. Jelinek R, Özkar S, Ozin GA (1992) J Am Chem Soc 114: 4907
52. Stein A, Ozin GA, Macdonald PM, Stucky GD, Jelinek R (1992) J Am Chem Soc 114: 5171
53. Jelinek R, Özkar S, Pastore HO, Malek A, Ozin GA (1993) J Am Chem Soc 115: 563
54. Engelhardt G, Koller H, Sieger P, Depmeier W, Samoson A (1992) Solid State Nucl Magn Reson 1: 127
55. Hunger M, Engelhardt G, Koller H, Weitkamp J (1993) Solid State Nucl Magn Reson 2: 111
55a. Verhulst HAM, Welters WJJ, Vorbeck G, van de Ven LJM, de Beer VHJ, van Santen RA, de Haan JW (1994) J Chem Soc, Chem Commun 639
56. Liang JJ, Sherriff BL (1993) Geochim Cosmochim Acta 57: 3885
57. Wu Y, Chmelka BF, Pines A, Davis ME, Grobet PJ, Jacobs PA (1990) Nature 346: 550
58. Jelinek R, Chmelka BF, Wu Y, Grandinetti PJ, Pines A, Barrie PJ, Klinowski J (1991) J Am Chem Soc 113: 4097
59. Grobet PJ, Samoson A, Geerts H, Martens JA, Jacobs PA (1991) J Phys Chem 95: 9620
60. Barrie PJ, Smith ME, Klinowski J (1991) Chem Phys Lett 180: 6
60a. He H, Klinowski J (1993) J Phys Chem 97: 10385
61. Chmelka BF, Wu Y, Jelinek R, Davis ME, Pines A (1991) In: Jacobs PA, Jaeger NI, Kubelková L, Wichterlová B (eds) Zeolite chemistry and catalysis. Elsevier, Amsterdam, p 435
62. Jelinek R, Chmelka BF, Wu Y, Davis ME, Ulan JG, Gronsky R, Pines A (1992) Catal. Lett. 15: 65
62a. Samoson A, Sarv P, van Braam Houckgeest JP, Kraushaar-Czarnetzki B (1993) Appl Magn Reson 4: 171
63. Rocha J, Klinowski J, Barrie PJ, Jelinek R, Pines A (1992) Solid State Nucl Magn Reson 1: 217
64. Peeters MPJ, de Haan JW, van de Ven LJM, van Hooff JHC (1992) J Chem Soc, Chem Commun 1560
65. Peeters MPJ, de Haan JW, van de Ven LJM, van Hooff JHC (1993) J Phys Chem 97: 5363
66. Peeters MPJ, van de Ven LJM, de Haan JW, van Hooff JHC (1993) J Phys Chem 97: 8254
67. Janchen J, Peeters MPJ, de Haan JW, van de Ven LJM, van Hooff JHC, Girnus I, Lohse U (1993) J Phys Chem 97: 12042
68. Engelhardt G, Sieger P, Felsche J (1993) Anal Chim Acta 283: 967
69. Ray GJ, Samoson A (1993) Zeolites 13: 410
70. Haddix GW, Narayana M, Gillespie WD, Georgellis MB, Wu Y (1994) J Am Chem Soc 116: 672
71. Janicke M, Kumar D, Stucky GD, Chmelka BF (1994) Stud Surf Sci Catal 84: 243
72. Haddix GW, Narayana M, Tang SC, Wu Y (1993) J Phys Chem 97: 4624
73. Xu Z, Sherriff BL (1993) Appl Magn Reson 4: 203; Teertstra DK, Sherriff BL, Xu Z, Cerný P (1994) Canad Mineralog 32: 69
73a. Smith ME, Jaeger C, Schoenhofer R, Steuernagel S (1994) Chem Phys Lett 219: 75
74. Bastow TJ, Hall JS, Smith ME, Steuernagel S (1994) Mater Lett 18: 197
75. Mueller KT (1991) PhD Dissertation, University of California, Berkeley; Lawrence Berkeley Laboratory Report No. LBL-31125
76. Mueller KT, Chingas GC, Pines A (1991) Rev Sci Instrum 62: 1445
77. Eastman MA, Grandinetti PJ, Lee YK, Pines A (1992) J Magn Reson 98: 333
78. Terao T, Fujii T, Onodera T, Saika A (1984) Chem Phys Lett 107: 145
79. Terao T, Miura H, Saika A (1986) J Chem Phys 85: 3816
80. Gerstein BC, Pan HJ, Pruski M (1990) Rev Sci Instrum 61: 2699
81. Mueller KT, Baltisberger JH, Wooten EW, Pines A (1992) J Phys Chem 96: 7001

82. Mueller KT, Wooten EW, Pines A (1991) J Magn Reson 92: 620
83. Baltisberger JH, Gann SL, Wooten EW, Chang TH, Mueller KT, Pines A (1992) J Am Chem Soc 114: 7489
84. Grandinetti PJ, Baltisberger JH, Llor A, Lee YK, Werner U, Eastman MA, Pines A (1993) J Magn Reson, Ser A 103: 72
85. Jeener J, Meier BH, Bachmann P, Ernst RR (1979) J Chem Phys 71: 4546
86. States DJ, Haberkorn RA, Ruben DJ (1982) J Magn Reson 48: 286
87. Youngman RE, Zwanziger JW (1994) J Non-Cryst Solids 168: 293
88. Farnan I, Grandinetti PJ, Baltisberger JH, Stebbins JF, Werner U, Eastman MA, Pines A (1992) Nature 358: 31
89. Pettifer RF, Dupree R, Farnan I, Sternberg U (1988) J Non-Cryst Solids 106: 408
90. Tossell JA, Lazzeretti P (1988) Phys Chem Miner 15: 564
91. Grandinetti PJ, Lee YK, Baltisberger JH, Sun BQ, Pines A (1993) J Magn Reson, Ser A 102: 195
92. Frydman L, Chingas GC, Lee YK, Grandinetti PJ, Eastman MA, Barrall GA, Pines A (1992) J Chem Phys 97: 4800
93. Frydman L, Lee YK, Emsley L, Chingas GC, Pines A (1993) J Am Chem Soc 115: 4825
94. Zwanziger JW, Olsen KK, Tagg SL (1993) Phys Rev B 47: 14618
95. Gann SL, Baltisberger JH, Pines A (1993) Chem Phys Lett 210: 405
96. Vega AJ (1992) J Magn Reson 96: 50
97. Vega AJ (1992) Solid State Nucl Magn Reson 1: 17
98. Baltisberger JH, Gann SL, Grandinetti PJ, Pines A (1994) Mol Phys 81: 1109
99. Gann SL, Baltisberger JH, Wooten EW, Zimmermann H, Pines A (1994) Bull Magn Reson 16: 68
100. Fyfe CA, Wong-Moon KC, Grondey H, Mueller KT (1994) J Phys Chem 98: 2139

Structural Studies of Noncrystalline Solids Using Solid State NMR. New Experimental Approaches and Results

Hellmut Eckert

Department of Chemistry, University of California, Santa Barbara, CA 93106, USA

Table of Contents

NMR Basic Principles and Progress, Vol. 33

Nuclear magnetic resonance is one of the most versatile spectroscopic tools for probing the local order of non-crystalline solids and glasses. In recent years, the research area has benefitted greatly from the influx of new powerful selective averaging and two-dimensional correlation techniques, resulting in significant progress in the structural understanding of the glassy state. The present review summarizes the state of the art in this research area. Applications to silicate, borate, phosphate and aluminum-containing oxide glasses are discussed, with an emphasis on most recent work utilizing more sophisticated techniques. The review also summarizes recent NMR results obtained in the area of covalent non-oxide glasses and their structural interpretation. Topics at the focus of interest include: (1) quantification of short-range and intermediate range order, (2) distribution functions for topological parameters (bond distances and angles) and (3) speciation equilibria and dynamics of the molten states.

1 Introduction

The term "glass" describes a state of matter that possesses most of the macroscopic and thermodynamic properties of a crystalline solid, while retaining the structural disorder and isotropic behavior typical of the liquid state. Cooling a liquid quickly below its freezing point under conditions that prevent thermodynamic equilibration results in the glassy state at a well-defined temperature T_g, where collective molecular motion is frozen abruptly. This "glass transition temperature" is a thermodynamic necessity and possesses the phenomenological appearance of a second order phase transition [1].

Due to the opportunities of fine-tuning the physical-chemical properties by composition and processing history, glasses have attained an ever increasing importance in materials science and technology over the last two decades [2]. In recent years, the quest for a fundamental understanding of such materials properties (and a desire to develop a rationale on how to control them) has spurred a plethora of structural studies [3]. While the complete geometric structure of a crystalline solid can be solved by a single X-ray or neutron diffraction study, for a glass there are no such experiments (or combinations thereof) that can result in information of comparable detail. Rather, one splits up the whole problem into a sub-set of individual structural questions, which are then studied by complementary techniques. Specifically, structural glass research deals with the following questions:

1. *Short-range order* (SRO) concerns the distribution of nearest-neighbor connectivities, coordination numbers and symmetries. From this (incomplete) point of view, the glass is an assembly of such microstructural units, to be identified and quantified.

2. *Intermediate-range order* (IRO) concerns second and third-nearest-neighbor correlations. IRO describes how the individual coordination polyhedra are linked to each other and which types of larger units, clusters, chains or rings can be identified in the glass.

3. *Topological disorder* describes the distribution of specific geometrical parameters. The lack of periodicity implies that certain interatomic distances and bond angles are not described by singular values but rather by distribution functions.

4. *Dynamic disorder* describes restricted atomic or molecular motion in the glassy state below T_g and the motional processes activated at the glass transition.

5. *The liquid state* bears much structural resemblance of the glassy state because it is the chemical equilibria present in the liquid state that are frozen in permanently at T_g. Thus, temperature dependent structural studies of the liquid state allow important conclusions on the structure of glasses.

6. *Macroscopic/general aspects:* Many glasses do not constitute homogeneous phases but are microphase-separated. The compositions and domain sizes of such microphases are an important part of the structural characterization. It is also of interest to know by which mechanism phase separation proceeds

(nucleation/growth or spinodal decomposition). Finally, the structural information obtained by various complementary techniques allows one to examine the applicability of a variety of general structural concepts of glass structure.

A wide variety of powerful structural approaches to the glassy state are widely employed, including X-ray and neutron diffraction, vibrational spectroscopy, EXAFS, Mossbauer spectroscopy, and solid state NMR. While the combined use of these techniques is indispensable for arriving at a comprehensive picture of a glass structure, it is also important that each technique be used to its fullest potential. This review has been written with the latter point in mind. Solid and liquid state NMR offer element-selective, inherently quantitative experimental approaches that have proven immensely powerful in the development of new and refined glass structure concepts. Furthermore, the selectivity of the NMR parameters specifically to the local structural environment of the nuclei under investigation ensures that the lack of periodicity in the glassy state does not lead to intolerable blurring of the spectroscopic responses, as is the case with diffraction experiments. NMR is thus particularly powerful for providing information about nearest-neighbor environments. Approximately 80% of all NMR studies of glasses published in the literature concern themselves primarily with the identification and quantification of such nearest-neighbor environments, primarily on the basis of wide-line and magic-angle spinning NMR lineshapes. Understanding the short range order is, however, only a first modest step towards developing a structural understanding of a glass. For this reason, this review will not attempt a detailed encyclopedic treatment of all the various isolated pieces of information that have become available in the countless glass systems studied to date. If desired, such details are available from a recent review [4]. Rather, this review will emphasize the process of how such short-range order information has been obtained, often with the help of non-standard NMR experiments, and how such information can be combined with theoretical studies and general condensed phase chemistry concepts to arrive at more refined glass structure models. Intermediate range order and site distribution effects are given special attention, where available, and throughout the review the emphasis will be on most recent results, particularly those not previously reviewed in Ref. [4].

2 Modern NMR Techniques Suitable for the Study of Glasses

Solid state NMR is useful for structural glass research because the spin Hamiltonian contains terms other than that describing the principal interaction of the nuclei with the external magnetic field [5–10]. These additional terms, which can be accounted for by standard perturbation theory, include direct (through-space) nuclear dipole-dipole coupling, H_d, indirect nuclear spin-spin coupling, H_J, chemical shielding, H_{CS} and for nuclei with spin quantum numbers

Table 1. Interactions in solid state NMR, spectroscopic parameters, and their selective measurement

Interaction	Hamiltonian	Parameters	Selective Measurement.	Ref.
Dipole-Dipole, homonuclear:	$H = \gamma^2\hbar^2 r_{ij}^{-3}(1-3\cos^2\theta)(3I_z^2\text{-}I^2)/2$ (two-spin Hamiltonian)	M_{2hom} or r_{ij}	90°-t-180° (spin echo decay)	40
	r_{IS} = internuclear distance	r_{ij}	Rotational Resonance	28
	θ = angle (r_{ij}, B_0)	r_{ij}	DRAMA	31
Dipole-Dipole, heteronuclear:		M_{2het} or r_{ij}	90°-t-180°(180°$_s$); SEDOR	43
	$H = \gamma_I\gamma_S\hbar^2 r_{IS}^{-3}(1-3\cos^2\theta)I_zS_z$ (two-spin Hamiltonian)	r_{IS}	Rotary Resonance Recoupling	29
		r_{IS}	Rotational Echo Double Resonance (REDOR)	30
	r_{IS} = internuclear distance θ = angle (r_{IS}, B_0)	T_{CR}	Cross-Polarization	32
Chemical Shift:	$H = \gamma I_z\sigma B_0$	$\delta_{11}, \delta_{22}, \delta_{33}$	MAS sideband analysis	
		$\delta_{11}, \delta_{22}, \delta_{33}$	2-D MAS-NMR (VACSY)	26
		δ_{iso}	MAS centerband position	15
			DAS, DOR, SATRAS (for $I > 1/2$)	18–20
Quadrupole:	$H = (e^2qQ)(3\cos^2\theta-1)(3I_z^2-I^2)/8I(2I-1)$ (for axial symmetry, $\eta = 0$)	e^2qQ/h, η	MAS lineshape analysis	16
	θ = angle (eq_{zz}, B_0)	(no easy separation)	DAS, DOR, SATRAS	18–20

$>1/2$, nuclear electric quadrupole coupling H_Q.

$$H_{pert} = H_d + H_J + H_{CS} + H_Q \tag{1}$$

Each of these interaction Hamiltonians can be written as the product of two tensors, which represent the spin part and the spatial part of the interaction in question. The spatial part contains information about the anisotropy of the perturbation and thus the structural and electronic environment of the nuclei under study. The NMR spectroscopist thus faces a dual task: (1) to accurately evaluate the Hamiltonian parameters (and their distribution functions) characterizing these internal interactions, and (2) to connect these Hamiltonian parameters with structural information.

All of the interactions in Eq. (1) are anisotropic, and H_J and H_{CS} also contain an isotropic part. Furthermore, H_d and H_J are comprised of a homonuclear contribution (arising from interactions with resonant nuclei) and heteronuclear contributions (arising from interactions with non-resonant nuclei). While the modification of resonance frequencies and solid state NMR lineshapes due to either H_{CS}, H_d, and H_J is readily calculated by first-order perturbation theory, the calculation becomes more complex if all of these internal interactions have to be considered simultaneously [11, 12], because the mutual tensor orientations must be known. Furthermore, the influence of H_Q upon the NMR lineshape frequently requires extension of the perturbation theory treatment to second order.

The most successful strategy for obtaining unambiguous information from NMR spectra is to simplify the effective spin Hamiltonian by "selective averaging" experiments. Table 1 summarizes those experiments that to date have become important in the study of glasses. Appropriate references have been included, since space limitations preclude a discussion of all the theoretical underpinnings. In the following the most common experiments with the greatest potential for applications to glasses are being discussed.

2.1 *Wideline NMR*

The simplest NMR experiment is the static measurement of the anisotropic NMR lineshape, using either continuous-wave (cw) or pulsed Fourier Transform spectroscopy. This approach has been often used in cases where the static lineshape is dominated by one type of interaction, hence making the use of selective averaging NMR experiments unnecessary. The most prominent examples are the low-field ^{11}B NMR spectra of the three-coordinate sites in borate glasses, whose lineshapes are dominated by second-order quadrupolar perturbations, all other effects being small. For spin-1/2 nuclei in glasses, wideline NMR powder patterns are sometimes useful for extracting anisotropic chemical shift information. Such spectra should be preferably taken with spin-echo sequences [13, 14] to avoid lineshape distortions due to signal loss during the receiver deadtime. Finally, the lineshapes of many metallic glasses are much wider than typical excitation bandwidths. In such cases, the whole static spectrum must be recorded by measuring the spin echo height as a function of transmitter frequency ("spin echo mapping").

2.2 Magic Angle Spinning NMR

The most frequently used selective averaging experiment involves recording a simple NMR spectrum while rotating the sample at an angle of 54.7 degrees (the "magic angle") relative to the magnetic field direction [15]. In the fast-spinning limit ($\omega_r \gg \omega_{int}$, ω_r being the rotor frequency and ω_{int} the resonance broadening arising from the internal interaction considered), this operation preserves only the isotropic parts of H_{pert}, while eliminating the effect of the anisotropic contributions to H_{CS}, H_J, as well as H_d and (to first order) H_Q, on the NMR lineshape. As a result, the spectroscopic resolution improves markedly, allowing the observation of (partially or fully) separated resonances for chemically inequivalent sites. Furthermore, for spin-1/2 nuclei the residual MAS-NMR line broadening observed in a glass reflects the distribution function of isotropic chemical shifts. In cases where there exists a theoretical or empirical relationship between chemical shift and structural parameters, this experimental chemical shift distribution can then be related to topological disorder effects in the glassy state. A prominent example is the correlation of ^{29}Si chemical shifts with average Si-O-Si bond angles. Based on this correlation, approximate bond angle distributions have been derived from the experimental ^{29}Si MAS-NMR peak shapes in glassy silica and other simple silicate glasses (see below).

2.3 Advanced Sample Reorientation Techniques

For quadrupolar nuclei ($I \geqq 1$), MAS only scales down rather than eliminates second order quadrupolar broadening effects [16]. As a result, the MAS-NMR lineshapes of glasses are influenced both by distributions in isotropic chemical shifts and nuclear electric quadrupole coupling parameters, and thus have limited structural significance. Most recently, more sophisticated sample re-orientation techniques have been developed, which eliminate the anisotropic broadening of MAS-NMR lineshapes due to second-order quadrupolar effects. Figure 1 shows a simple version of the dynamic angle spinning (DAS) experiment [17]. Following a 90° preparation pulse, the sample spins initially at angle θ_1 for a time t_1. The phase evolution, which depends on the value of the angle and the quadrupolar Hamiltonian is subsequently stored along the z-direction by a second 90° pulse, and the spinning axis is flipped into the orientation θ_2. The third 90° pulse initiates phase evolution under the quadrupolar Hamiltonian modulated by spinning at θ_2. The angles θ_1 and θ_2 are chosen such that evolution following the third pulse reverses that following the first pulse. In the simplest case when $\theta_1 = 37.38°$ and $\theta_2 = 79.19°$ refocusing occurs at time t_1 following

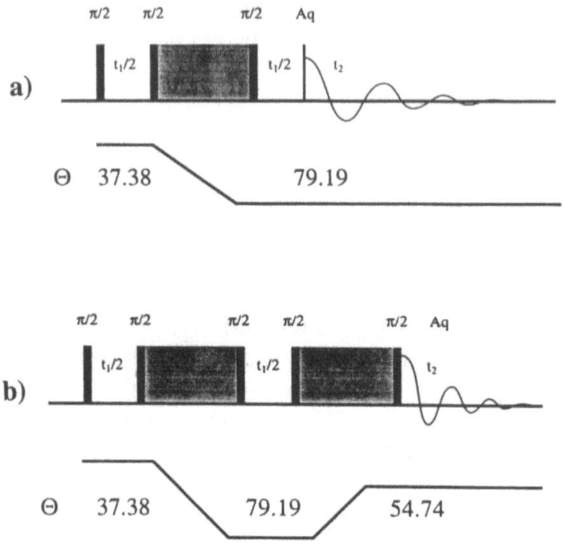

Fig. 1a, b. Pulse sequence used in the DAS experiment **a** simplest two-angle version, **b** with detection at the magic angle. During the period characterized by the *shaded areas* the rotor angle is flipped. Reproduced with permission from Ref. [17c]

the third pulse, resulting in formation of an echo. Incrementation of t_1 then allows measurement of evolution exclusively under the isotropic chemical shift. Consequently, well-resolved peaks are observable, appearing at the resonance position

$$\delta_{\mathbf{exp}} = \delta_{\mathbf{iso}} + \delta^{(2)} \tag{2}$$

where $\delta^{(2)}$ is the second-order quadrupole shift. For the central transition of half-integer quadrupolar nuclei $\delta^{(2)}$ is calculated according to [18]:

$$\delta^{(2)} = -(3/40)\nu_0^{-2}(e^2qQ/h)^2[I(I+1)-3/4]/I^2(2I-1)^2 \tag{3}$$

The same effect can be accomplished by simultaneous sample reorientation about the magic angle and a second angle inclined by 30.6 degrees relative to the MAS axis. This technique, known as double rotation (DOR) [19] is illustrated in Fig. 2.

2.4 2-D Chemical Shift Correlation Spectroscopy

Correlation spectroscopy (COSY) aims at uncovering connectivity between sites with different resonance frequencies (chemical shifts). This relationship is detected as coherence transfer that is mediated by scalar coupling. Figure 3a shows the pulse sequence. Following the initial 90° preparation pulse, the spins are frequency labeled during the evolution time t_1 according to their individual chemical shifts. The second 90° pulse generates coherence transfer among coupled spins, resulting in an altered precession frequency during the detection

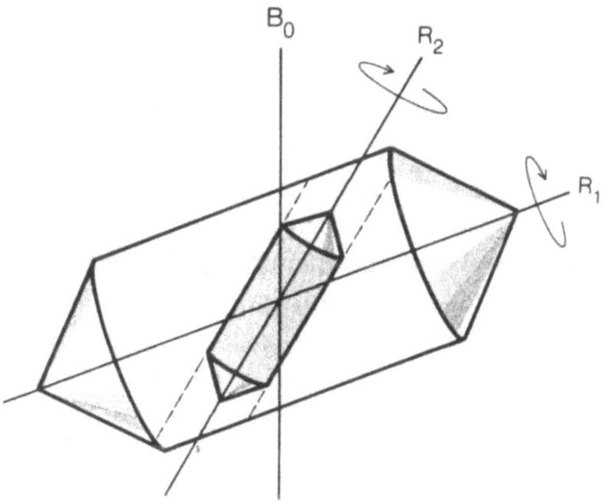

Fig. 2. Sample reorientation using a double rotor. The outer rotor spins at the magic angle (54.7°), while the axis of the inner rotor is inclined by 30.6° relative to the outer rotor axis. This sample reorientation leads to the cancellation of anisotropic second-order quadrupolar broadening, which is proportional to the second and fourth Legendre polynomials. Reproduced with permission from Ref. [19]

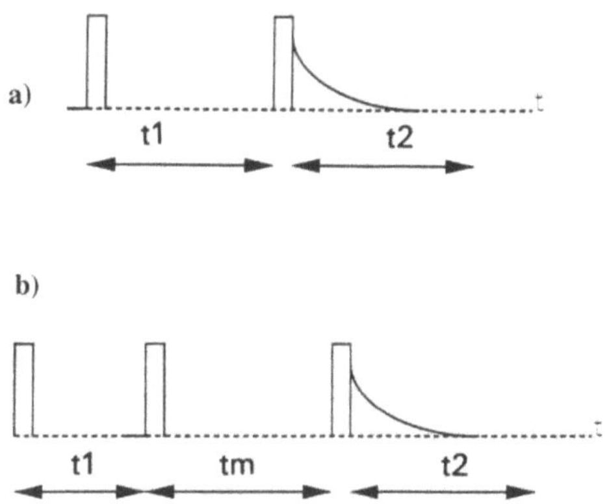

Fig. 3a, b. Pulse sequences for two-dimensional correlation spectroscopy: **a** COSY (correlation established by scalar couplings) **b** NOESY (correlation established by dipolar couplings or chemical exchange). The evolution time t_1 is systematically incremented. During the mixing period t_m coherence transfer occurs via chemical exchange or spin diffusion. Double Fourier Transform with respect to both t_1 and t_2 yields the 2-D NMR chemical shift correlation plot

period t_2. Double Fourier Transform with respect to t_1 and t_2 then affords the two-dimensional correlation plot, in which the presence of scalar couplings among spins manifests itself in the form of crosspeaks linking the isotropic chemical shifts δ_1 and δ_2 of the interconnected sites. While the COSY technique has been applied to liquids ever since its inception [20], applications on rotating crystalline solids have been much more recent [21], and Knight et al. were the first to apply it to glasses [22] (see below).

Figure 3b shows a similar kind of experiment, in which the coherence transfer occurs as a result of spatial or spin diffusion, allowed to proceed during a fixed time t_m. In recent years, this pulse sequence has been used to characterize chemical exchange occurring in glasses above the glass transition temperature as a result of bond-breaking/bond making processes [23, 24].

2.5 2-D Isotropic/Anisotropic Correlation Spectroscopy

A frequent dilemma characterizing standard MAS-NMR spectra of glasses is the poor resolution due to wide isotropic chemical shift distributions, resulting essentially in just a single broad resonance without any chemical information. In principle, the resolution of such spectra can be improved if there is a clear correlation between isotropic chemical shift and another anisotropic parameter, such as the nuclear electric quadrupolar interaction, the chemical shift anisotropy or the magnetic dipole-dipole coupling constant. Such information is implicitly provided by the two-dimensional DAS experiment, where the isotropic chemical shift is correlated with the second-order quadrupolar broadening in the anisotropic dimension [17a]. Farnan et al. have used this separation to elucidate bond angle distributions in silicate glasses [25].

Frydman et al. describe another elegant experiment, which is equally applicable to spin-1/2 nuclei [26]. Here the isotropic and anisotropic interactions are separated in the two dimensions by processing signals arising from a spinning sample, acquired in independent experiments as a function of the angle between the spinning axis and the magnetic field. This method results in a two dimensional correlation of isotropic chemical shifts with chemical shift anisotropies and has been applied recently to glasses by Zwanziger and coworkers [27].

2.6 MAS-NMR with Dipolar Editing

The selective averaging accomplished in MAS NMR is rather drastic. For spin-1/2 nuclei in the fast-spinning limit all of the anisotropic information is eliminated from the spectrum. To enhance the informational content of regular MAS NMR, it is often possible to re-introduce anisotropies (in general dipolar couplings) in a controlled fashion during the experiment for spectral editing purposes. Thereby the existence of homo- and heteronuclear dipole-dipole or scalar couplings can be proved and hence serve to detect connectivity or spatial proximity among sites resolved in MAS-NMR. Broadly speaking, the techniques used for this objective can be grouped into recoupling and coherence transfer techniques. *Recoupling techniques*, such as rotational resonance (R^2) [28], rotary resonance recoupling (R^3) [29], Rotary Echo Double Resonance (REDOR) [30], and Dipolar Recovery At the Magic Angle (DRAMA) [31] re-introduce the homo- or heteronuclear dipole-dipole couplings normally eliminated by magic angle spinning to alter the spectral appearance of specific sites and to measure internuclear distances. Figure 4 shows a simple version of the REDOR experiment. On the other hand, *coherence transfer techniques* such as cross-polarization (CP) [32] (Fig. 5), transferred echo double resonance (TEDOR) [33], spin diffusion NMR [34, 35] (Fig. 6), and multiple-quantum-filtered MAS NMR [36] create special homo- or heteronuclear coherences between select sites due to the presence of dipole-dipole or scalar couplings.

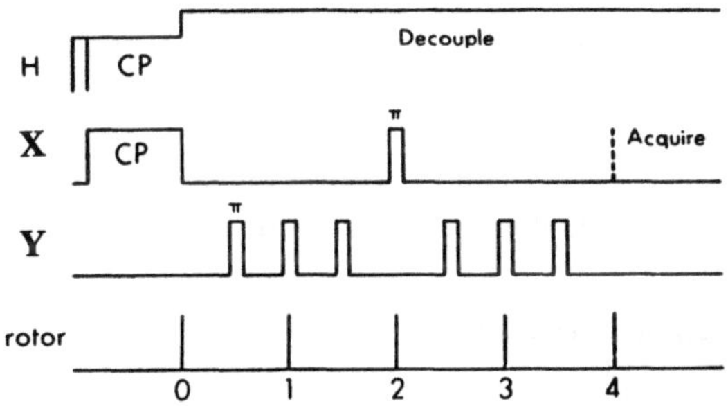

Fig. 4. Pulse sequence used in the REDOR experiment for the example of measuring X–Y dipolar coupling. In this case the preparation of transverse coherence is accomplished by ^1H–X cross-polarization (see Fig. 5). The 180° pulses on the Y channel re-introduce the dipole-dipole coupling between nuclei X and Y normally averaged under MAS conditions. Reproduced with permission from Ref. [30b]

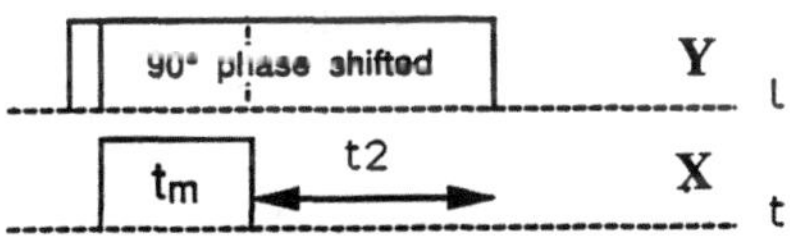

Fig. 5. Pulse sequence for cross-polarization. Following an initial 90° pulse the Y magnetization is spin locked along an applied B_1 field. For the duration of the mixing time t_m a long pulse is supplied to the X spins such that the X and Y precession frequencies in the rotating frame are equal. Subsequently, the transverse X magnetization is acquired during the acquisition period t_2, while the Y spins are continuously irradiated to provide decoupling

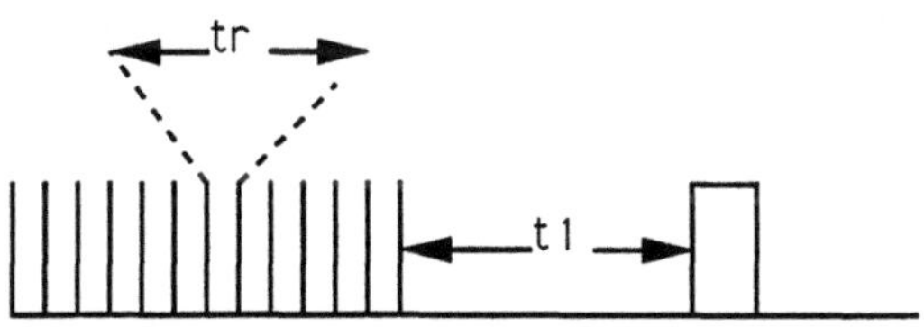

Fig. 6. 1-D spin diffusion MAS experiments, using a rotor-synchronized DANTE sequence for selective inversion of a specific resonance. The sum of the flip angles separated by the rotor period t_r equals a 180° pulse. Following the incremented mixing period t_1, all spins are affected with a non-selective 90° detection pulse

Perhaps the most widely applicable experiment within this group is the CPMAS technique, which achieves magnetization transfer from an abundant-spin system (usually protons) to the observe-nuclei whose signal is desired. Recently, there has been substantial progress in exploiting CPMAS from nuclei other than ^1H to enhance the detection sensitivity of rare nuclei. [37] This progress is expected to produce significant new applications to inorganic glasses, which are usually devoid of protons.

2.7 *MAS-NMR on Quadrupolar Satellite Transitions ("SATRAS")*

As shown by Samoson [19], the broadening $\Delta(m)$ and line displacement $\delta^{(2)}$ caused by second order quadrupolar effects depend on the transition quantum number according to:

$$\Delta(m) = \frac{3}{128}\frac{C_Q^2}{\nu_0^2}\frac{6I(I+1) - 34m(m-1) - 13}{I^2(2I-1)^2} \tag{4}$$

$$\delta^{(2)} = \frac{3}{40}\frac{C_Q^2}{\nu_0^2}\frac{I(I+1) - 9m(m-1) - 3}{I^2(2I-1)^2}\left(1 + \frac{1}{3}\eta^2\right) \tag{5}$$

One of the most important consequences of these formulae is that for quadrupolar spin-5/2 nuclei the second-order quadrupole effects on the first pair of quadrupolar satellite transitions ($m = 3/2 \rightarrow m = 1/2$ and $m = -1/2 \rightarrow m = -3/2$) are minimized [38]. Specifically Eqs. (4) and (5) predict:

$$\delta^{(2)}_{3/2}/\delta^{(2)}_{1/2} = -0.1235 \tag{5a}$$

and

$$\Delta_{3/2}/\Delta_{1/2} = 0.2917 \tag{4a}$$

Thus, for such nuclei substantially better resolution between individual sites in glasses is expected in the satellite transitions than in the central transitions. Furthermore, the chemical shifts and their distributions associated with such sites can be measured much more accurately. Experimentally, satellite transition spectroscopy ("SATRAS") is generally carried out under MAS conditions at fairly high spinning speeds ($\geqq 10$ kHz), which generate a spinning sideband manifold from the quadrupolar satellite powder pattern (see Fig. 7). The positions

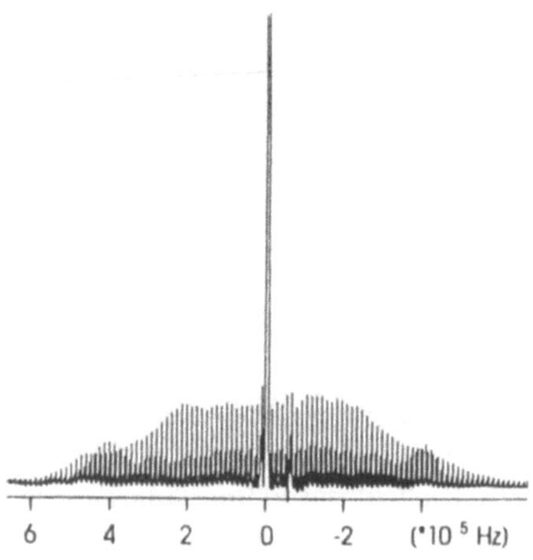

Fig. 7. Typical spinning sideband manifold generated from quadrupolar satellite powder pattern of the compound $9Al_2O_3 \cdot 2B_2O_3$. Reproduced with permission from Ref. [39]

of these sidebands enable precise measurements of isotropic chemical shifts, and their lineshapes convey information about the isotropic chemical shift distribution. For precise quantitative site information the overall sideband pattern must be analyzed, requiring (a) extremely stable spinning speeds, (b) detailed attention to baseline artifacts, and (c) a uniform excitation over the entire spinning sideband manifold, or a mathematical correction if the latter cannot be achieved [39].

2.8 Static Spin Echo and Spin Echo Double Resonance NMR

Static spin echo NMR experiments are useful for selectively extracting information about internuclear dipole-dipole couplings within multispin systems. Homonuclear interactions can be measured by a simple Hahn spin echo (90°-t_1-180°) pulse sequence [40] (Fig. 8a). Under certain assumptions, for a system characterized by multiple interactions the decay of the spin echo intensity $I(2t_1)$ as a function of the variable pulse delay t_1 is Gaussian, and affords the measurement of an average second moment M_2 characterizing the strength of the dipole–dipole interactions among resonant nuclei. In a glass, a distribution of second moments is present, resulting in a distribution of Gaussian decay curves:

$$I(2t_1)/I(0) = \sum \exp^{-}\{(2t_1)^2 M_2/2\} \tag{6}$$

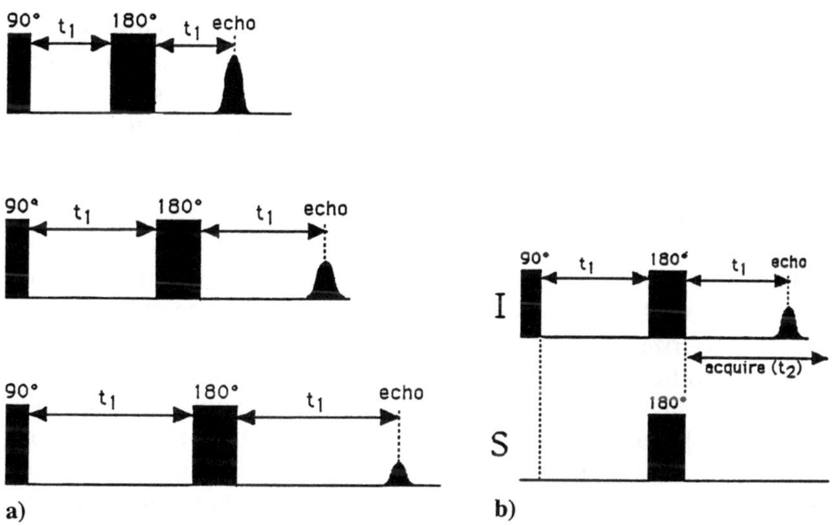

Fig. 8a, b. Pulse sequences for the measurement of homo-and heteronuclear dipole-dipole coupling constants using **a** spin echo and **b** spin-echo double resonance (SEDOR) techniques. Evolution times t_1 are systematically incremented

Reimer and Duncan were the first to probe homonuclear dipole–dipole couplings in amorphous substances using spin echo NMR [41]. The technique was used subsequently to investigate the distribution of P and F in glassy silica [42a,b], and, more recently, to characterize quantitatively the distribution of ^{31}P spins in a variety of non-oxide glasses [4,42c].

The heteronuclear version of this experiment, spin-echo double resonance (SEDOR) (Fig. 8b), developed and applied to catalytic materials by Slichter and coworkers [43], works in a similar fashion. Here the 180° pulse applied to the (unobserved) Y-spins during the evolution period interferes with the refocusing of the X–Y dipole–dipole coupling by the spin echo sequence. Incrementation of the evolution time t_1 with and without the 180°_Y pulse affords the selective measurement of this interaction. The SEDOR experiment has been useful for characterizing heterodipolar couplings in doped amorphous silicon materials [44] and non-oxide glasses [45]. Both the static spin echo and the SEDOR experiments are structurally very valuable, since homo- and heterodipolar second moments can be calculated readily from the van Vleck equation [46]:

$$M_2 = (3/5)(\mu_0/4\pi)^2\gamma^4\hbar^2 I(I+1)\sum r_{ij}^{-6} \quad \text{(homo)} \tag{7a}$$

$$M_2 = 4/15(\mu_0/4\pi)^2\gamma_I^2\gamma_S^2\hbar^2 S(S+1)\sum r_{IS}^{-6} \quad \text{(hetero)} \tag{7b}$$

Comparison of experimental and simulated spin echo or SEDOR decay curves then serves to develop and test hypothetical atomic distribution models in glasses. It must be emphasized that both techniques have to be applied judiciously, and attention must be paid to the detailed spin dynamics present in the materials under study. For instance, spin echo and SEDOR NMR cannot differentiate between direct and indirect dipole–dipole interactions. Since the latter cannot be calculated from structural information the usefulness of these experiments is restricted to cases where $\mathscr{H}_d \gg \mathscr{H}_J$. A more detailed discussion of the premises of this technique, along with experimental results on model compounds has been given [47]. In practice, spin-echo techniques have proven quite successful in the structural analysis of phosphorus-based systems [4].

2.9 Multiple-Quantum NMR

Multiple Quantum NMR can be used to characterize clustered or homogeneous distributions of spin-1/2 nuclei in solids on the 10–20 Å length scale. The technique characterizes the buildup of multiple-quantum coherences as a function of preparation time. (see Fig. 9). While the focus of earlier work was to determine the highest-order coherence for counting the number of nuclei in a given cluster [48], more recent studies have focused on the detailed characterization of the dynamics of multiple-quantum coherences, and their correlation with the strength of the internuclear dipole-dipole couplings and the dimensionality of the spin distribution [49]. Figure 10 shows that scaling of the preparation time by the second moment describing the strength of the internuclear dipolar coupling results in rather uniform behavior for different types

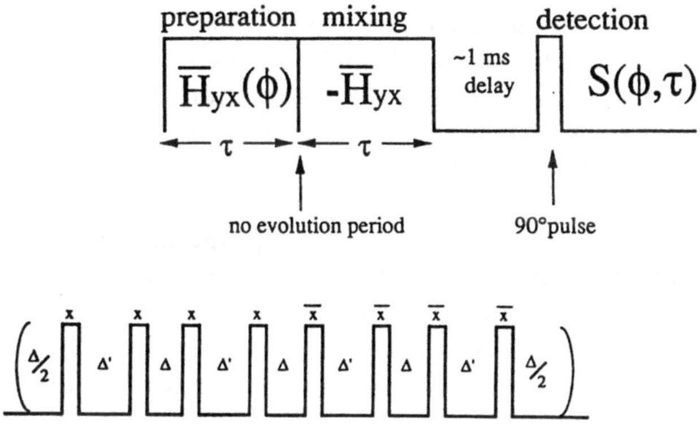

Fig. 9. Timing diagram for the even-order selective MQ experiment. The signal $S(\phi, t)$ is detected as the length of the preparation period and its phase relative to the mixing period, ϕ, are varied. The bottom part of the figure shows a typical preparation period. Reproduced with permission from Ref. [49b]

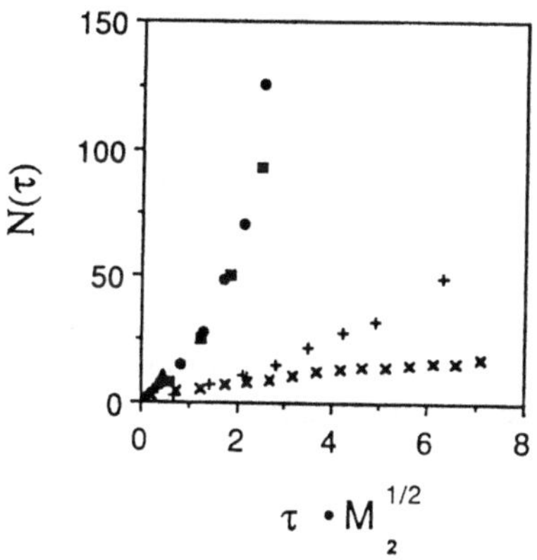

Fig. 10. Effective number of correlated spins, $N(\tau)$ versus preparation time τ, scaled by the square root of the second moment characterizing the dipole-dipole coupling. For the three 3-D systems Ca^1H_2, $Ca^{19}F_2$, and $Ga^{31}P$ (*circles, squares,* and *triangles*) universal behavior is observed. +: data for an infinite 2-D system, hydrogenated diamond powder, x: a finite 21 spin system. Reproduced with permission from Ref. [49b]

of 3-D solids, and allows distinction between 1-D, 2-D, and 3-D spin distributions. Although to date no applications of this approach to inorganic glasses have been reported, the technique appears to be promising for such applications.

3 Silicate Glasses

Amorphous silica forms a continuous random network (CRN), where $SiO_{4/2}$ tetrahedra are linked by corner-sharing. The addition of alkali or alkaline-earth oxides to the silica melt depolymerizes this network resulting in the formation of silicon species bonded to non-bridging oxygen atoms. The distribution of the non-bridging oxygen atoms in this modified random network (MRN) has been of considerable interest, and is related to the distribution effects of the modifier cations. Furthermore, the silicate network is altered by the simultaneous presence of other network formers, by intermediate oxides such as aluminum and lead oxide, and by volatile constituents such as water and carbon dioxide. Fairly recently, it has also been shown that the structural properties of silicate glasses can be substantially modified via anion substitution with fluoride, carbide, and nitride species. The structural implications of such compositional modifications are of great interest, and have been widely studied in recent years by modern solid state NMR techniques [4]. The present review will be largely limited to new developments in this area.

3.1 Bond Ordering in Binary Alkali Silicate Glasses

The modified random network (MRN) of simple binary alkali and alkaline earth silicate glasses is generally understood in terms of a partially depolymerized $SiO_{4/2}$ network, containing the modifier ions in its interstices. Recent neutron diffraction data of $CaSiO_3$ glass have shown that these cationic species are not randomly distributed over the network, but rather are located in fairly well-defined structural environments [50]. Furthermore, there are well-defined cation-cation distance correlations, signifying the presence of substantial intermediate range order. This is not unexpected in view of the known structural data on crystalline silicates. Based on such information it is suggested that in silicate glasses each modifier cation is coordinated to 4–6 non-bridging oxygen atoms (nbo-s), and likewise, that each nbo is coordinated by 4–6 modifier ions [51]. Such a model then raises the question of bond ordering, both in a *chemical* and in a *spatial* sense.

3.1.1 Chemical Bond Ordering

Bond ordering in a *chemical* sense addresses the quantitative distribution of the non-bridging oxygen atoms over the silicon atoms. In the MRN model, five individual types of $Si^{(n)}$ units are possible, depending on the number n of bridging oxygen atoms attached ($4 \geqq n \geqq 0$). Two widely discussed distribution models are known in the literature as the ordered ("binary") model and the random model, respectively. The *ordered* model, which is essentially an extension of the situation in crystalline silicates, minimizes the number of $Si^{(n)}$ species at any given composition. Thus, in "stoichiometric" glasses with compositions $M_2Si_2O_5$ ($n = 3$), M_2SiO_3 ($n = 2$), $M_6Si_2O_7$ ($n = 1$), and M_4SiO_4 ($n = 0$), only the corresponding $Si^{(n)}$ species is present, whereas "nonstoichiometric" glasses only contain those $Si^{(n)}$ species attributed to the two bordering stoichiometric compositions. In contrast, the *random* model maximizes the configurational entropy, and all four $Si^{(n)}$ species are present at all compositions to varying extents, following a binomial distribution law.

To clarify this issue, a large body of ^{29}Si MAS-NMR data exists now in the literature on a wide range of compositionally diverse glassses [4]. As widely documented in crystalline systems, the five $Si^{(n)}$ sites are distinguishable by their ^{29}Si isotropic chemical shifts [52]. Furthermore, the spherically symmetric $Si^{(0)}$ and $Si^{(4)}$ units, which lack chemical shift anisotropy broadening, yield sharp resonances in static ^{29}Si NMR, and are thereby easily distinguished from the anisotropically broadened $Si^{(3)}$, $Si^{(2)}$, and $Si^{(1)}$ resonances [53]. The basis for the spectroscopic separation is summarized in Fig. 11. For binary alkali silicate glasses the spectral resolution in the MAS-NMR spectra is sufficiently good, and $Si^{(n)}$ species distributions have been derived from the spectra by deconvolution [54]. Typical results are shown in Fig. 12. The results obtained in this fashion show behavior that is generally in-between the predictions made by the

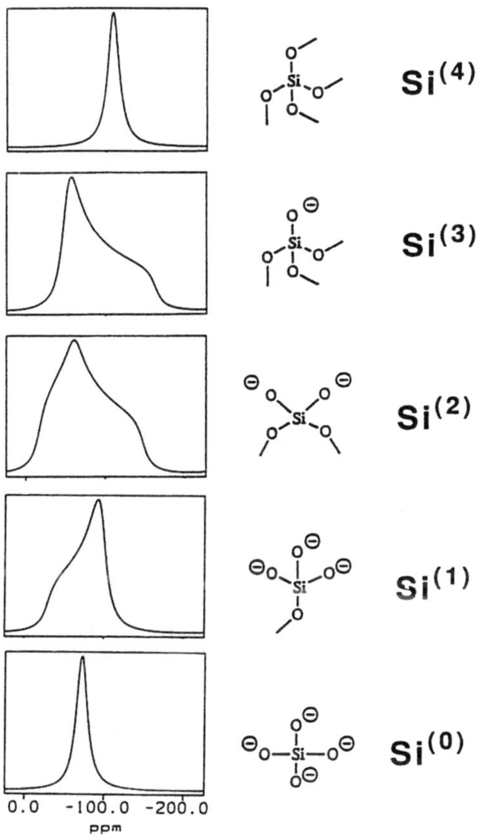

Fig. 11. $Si^{(n)}$ (also labeled $Q^{(n)}$ in the literature) species in alkali silicate glasses and their ^{29}Si chemical shift powder patterns

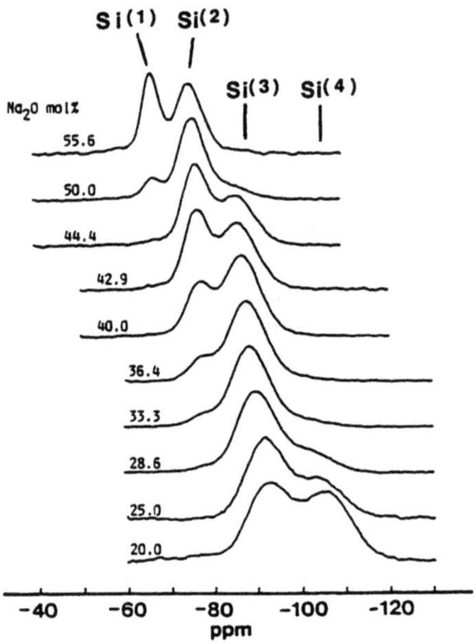

Fig. 12. Compositional dependence of the ^{29}Si MAS-NMR spectra in the glass system Na_2O–SiO_2. Peak assignments are indicated. Reproduced with permission from Ref. [55]

two models. A generalized description of this intermediate situation utilizes disproportionation reactions of the kind [51, 55]:

$$Si^{(3)} \Leftrightarrow Si^{(2)} + Si^{(4)}; \quad Si^{(2)} \Leftrightarrow Si^{(3)} + Si^{(1)}; \quad Si^{(1)} \Leftrightarrow Si^{(2)} + Si^{(0)}$$

characterized by equilibrium constants K_3, K_2, and K_1, respectively. $K_i = 0$ corresponds to the ordered model, $K_i = 1$ describes the random situation, while $K_i \gg 1$ would eventually result in segregation and crystallization phenomena. A survey of the data shows that in general K_i increases with increasing polarizing power Z/r of the cation in the glass: K_i (Li) > K_i(Na) > K_i(K). For potassium silicate glasses K_i is so small that the distribution approaches that predicted by the binary model [56]. Furthermore, in general $K_1 > K_2 > K_3$ [51]. The effect of the alkali ions upon the silicon site distribution may reflect changes in the electronic properties of the Si—O bonds. which have been suggested by recent ab initio MO calculations [57]. Finally, in glasses with high modifier contents (>40 mol%), the number of non-bridging oxygen atoms as found from the experimental $Si^{(n)}$ speciations is sometimes lower than expected based on the alkali content, hence suggesting the presence of free oxide (O^{2-}) ions. This assertion remains to be confirmed, however, by ^{17}O DAS or DOR-NMR studies.

Attempts of extending the ^{29}Si NMR analysis of $Si^{(n)}$ units to alkaline-earth silicate glasses and compositionally more complex systems have proven less successful, because in these systems the chemical shift dispersion for a given $Si^{(n)}$ species is comparable to the chemical shift differences between different

$Si^{(n)}$ sites [58]. As a result, the ^{29}Si MAS-NMR spectra are poorly resolved, hence precluding a meaningful analysis.

3.1.2 Spatial Bond Ordering

Understanding the detailed $Si^{(n)}$ species distribution in a silicate glass is only a first step towards developing an overall understanding of glass structure. As discussed above, in alkali silicate glasses the non-bridging oxygen atoms tend to aggregate around modifier cations and vice versa, resulting in a cation distribution that resembles clustering at least at low modifier concentrations. This raises the question of the spatial $Si^{(n)}$—O—$Si^{(n)}$ connectivity. The question is whether all of the various $Si^{(n)}$ units can be viewed as part of a single extended network or whether units such as $Si^{(3)}$ or $Si^{(2)}$ are mostly connected among themselves, thus forming small island clusters rich in cations within the glass.

It has been reported that, for a given glass system, the chemical shifts attributed to a given $Si^{(n)}$ unit change systematically with the composition of the glass, thus supporting the extended network concept [54a]. All $Si^{(n)}$ resonances move consistently downfield with increasing modifier content, presumably reflecting the increased average concentration of bridging oxygen atoms in the third coordination sphere as the network is depolymerized. More direct evidence for direct $Si^{(n)}$—O—$Si^{(n-1)}$ connectivity has come from two-dimensional COSY-MAS-NMR studies (see Fig. 13) on two ^{29}Si enriched sodium silicate glasses with Na_2O contents of 20 and 41 mol%, respectively [22]. The cross-peaks seen in Fig. 13 between the $Si^{(3)}$ and $Si^{(4)}$ resonances arise from coherence transfer via two-bond scalar coupling, thus providing direct evidence for such interconnections. Surprisingly, participation in a $Si^{(n)}$—O—$Si^{(n-1)}$ linkage appears

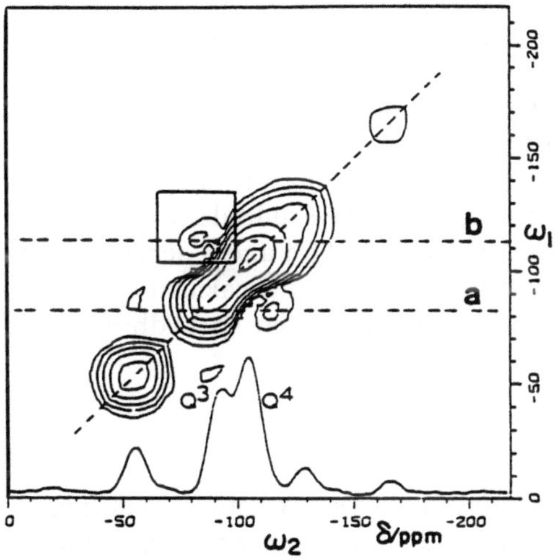

Fig. 13. Two dimensional ^{29}Si COSY spectroscopy of a glass with composition $Na_2Si_4O_9$. The $Si^{(3)}$–$Si^{(4)}$ connectivities are revealed by the crosspeaks *b* and *a*. Reproduced with permission from Ref. [22]

to shift the $Si^{(n)}$ resonance further upfield and the $Si^{(n-1)}$ resonance further downfield as compared to respective majority sites. This finding is unexpected in view of the chemical shift trends described above and possibly signifies bond angle redistributions that may be unique to such connectivities. Furthermore, these 2-D experiments only signify that such interconnections exist in principle, but do not allow one to judge easily how abundant they may be. The quantitative determination of such $Si^{(n)}$—O—$Si^{(n-1)}$ links thus remains a challenging subject for future studies.

3.2 Bond Angle Distribution in Silicate Glasses

It is generally accepted that the nearest neighbor environments in silicate glasses are rather well-defined and that the intrinsic disorder arises from variations in Si—O—Si bond angles. Thus, to characterize this disorder, one wishes to evaluate this distribution function as accurately as possible. This has been attempted by ^{29}Si MAS NMR. The spectra of glassy SiO_2 and silicate glasses are exclusively influenced by distributions of isotropic chemical shifts. Thus, numerous authors have attempted to "translate" the ^{29}Si chemical shift dispersion into the Si—O—Si bond angle distribution function using various semi-empirical correlations of ^{29}Si chemical shifts with Si—O—Si bond angles (for details see [4] and references therein). The resulting bond angle distributions are rather similar with maxima between 141° and 145° for glassy SiO_2, but substantially narrower than deduced on the basis of diffraction data. Dupree and coworkers have pointed out, however, that neither of these correlations succeed in simulating the experimental spectrum of tridymite, a SiO_2 polymorph containing six crystallographically different $SiO_{4/2}$ units, accurately [59]. Thus, at present it is not clear which of these chemical shift correlations is the "best" one. A semi-empirical bond polarization theory has been introduced, which for $Si^{(4)}$ sites accounts extremely well for the small differences in chemical shifts observed as a function of bond angles [60]. However, a fundamental difficulty encountered with ^{29}Si chemical shift/bond angle correlations is the fact that the measured chemical shift reflects the composite effect of four Si—O—Si bond angles. For this reason, ^{17}O NMR parameters associated with the bridging oxygen atoms are expected to give more direct information. Ab initio calculations carried out on the model molecule H_3Si—O—SiH_3 have indeed shown that ^{17}O nuclear electric quadrupolar coupling parameters e^2qQ/h and η depend on the intertetrahedral bond angles α [61]. As a matter of fact, the calculations reflect the same trend as predicted by the rather simple oxygen sp hydbridization picture:

$$e^2qQ/h \sim \cos\alpha/(1-\cos\alpha) \tag{8}$$

$$\eta = 1 + \cos\alpha \tag{9}$$

Thus, with increasing α value, e^2qQ/h increases, whereas η decreases, approaching the value of zero for the cylindrically symmetric environment when

$\alpha = 180°$. These correlations are qualitatively confirmed by experimental ^{17}O NMR measurements on the compounds cristobalite and wadeite [62].

Obtaining a distribution of Si—O—Si bond angles α then requires that the distribution functions of e^2qQ/h and/or η be determined in an accurate fashion from the ^{17}O NMR lineshape parameters associated with the bridging oxygen atoms. The first attempt in this direction was made by Geissberger and Bray who analyzed the low-field ^{17}O continuous wave NMR spectrum of SiO_2 glass in this fashion [63]. More recently, Farnan et al. have published a very elegant study on glassy $K_2Si_4O_9$, using DAS NMR, which results in a completely isotropic spectrum in the projection on the ω_1 axis [25]. When applied to $K_2Si_4O_9$ glass, the resonances due to bridging and non-bridging oxygen atoms at 9.4 T are perfectly resolved, permitting detailed analysis of the bridging oxygen resonance.

Since the individual slices in the ω_2 domain of this experiment still display the characteristic lineshapes due to second-order quadrupolar perturbations,

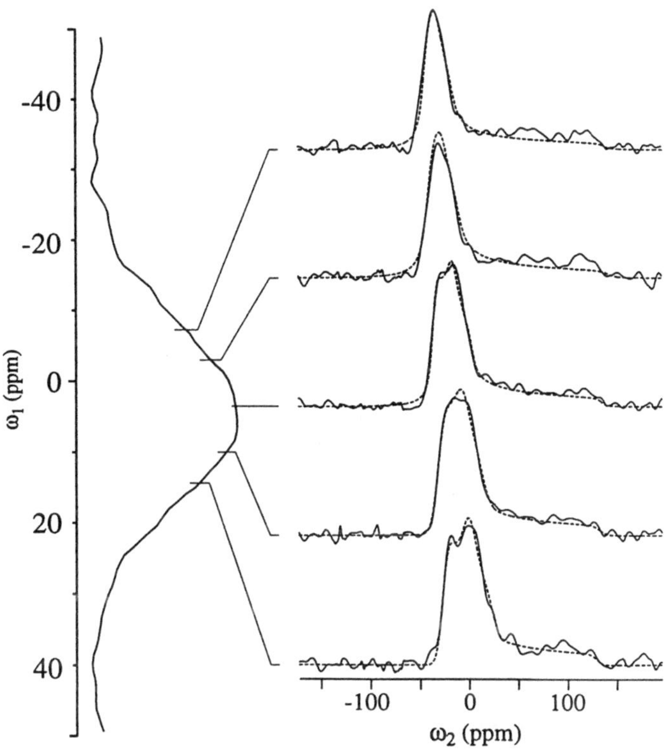

Fig. 14. ^{17}O DAS-NMR results on glassy $K_2Si_4O_9$. Indicated are the individual anisotropic slices for various points across the broadened DAS spectrum in the isotropic dimension. Note the systematic change in the asymmetry parameter. *Dashed curves* are simulated spectra. Reproduced with permission from Ref. [25]

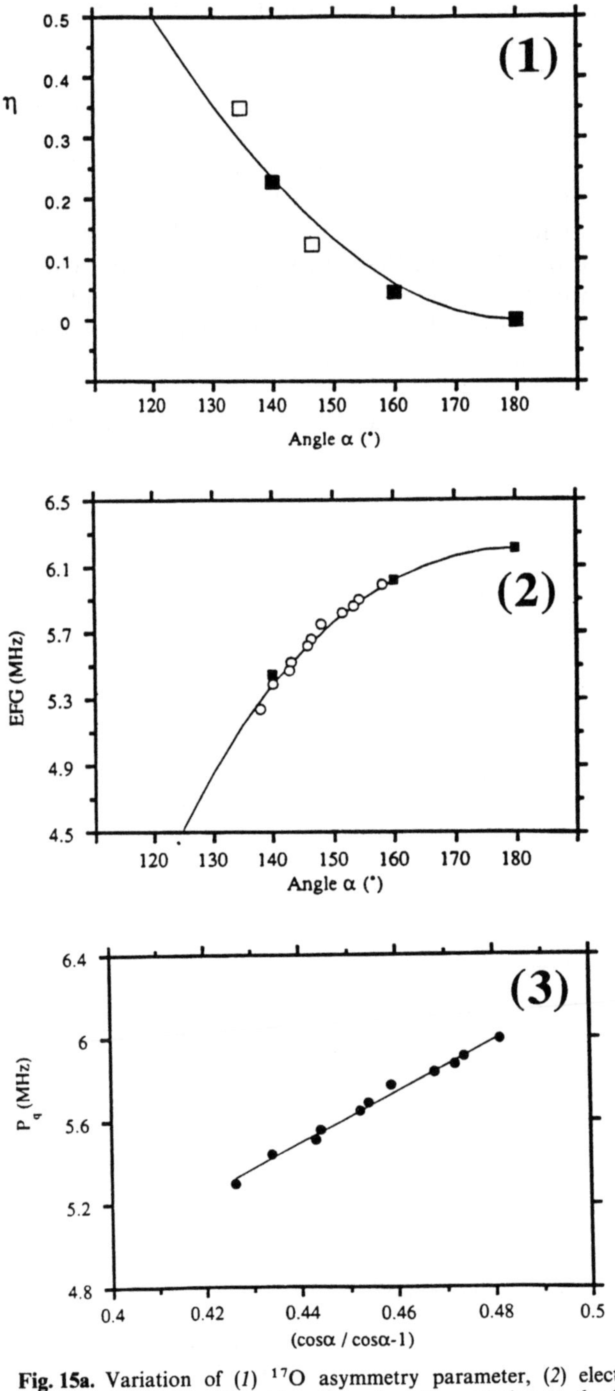

Fig. 15a. Variation of (*1*) ^{17}O asymmetry parameter, (*2*) electric field gradient, and (*3*) the quadrupolar product P_q with various functions relating to the Si—O—Si bond angle α. *Filled squares* are results from ab-initio calculations. Other details are given in Ref. [25]

the DAS technique permits interesting correlations of isotropic and anisotropic lineshape parameters in this case. Figures 14 and 15a reveal that the magnitude and asymmetry of the electric field gradient (as determined from computer fitting of the ω_2 slices) varies systematically across the isotropic lineshape in the ω_1 domain: Progressing from low to high values of ω_1, e^2qQ decreases while η increases, in good agreement with the predictions made by Eqs. (8) and

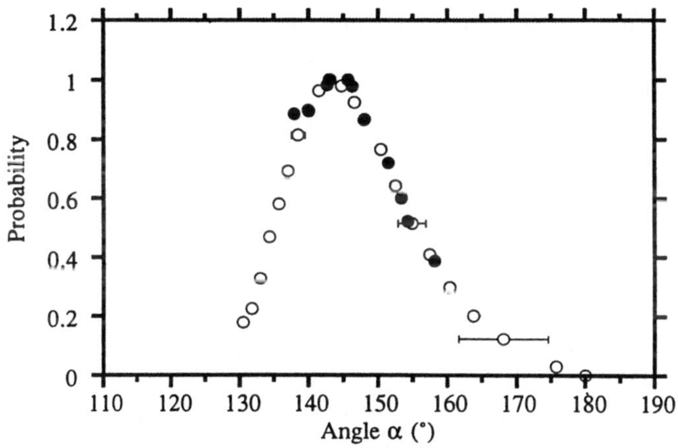

Fig. 15b. Si—O—Si bond angle distribution function derived from the DAS-NMR data (○) from the fitted η values and the correlation in Fig. 15a. (●) using the procedure described in the text. Reproduced with permission from Ref. [25]

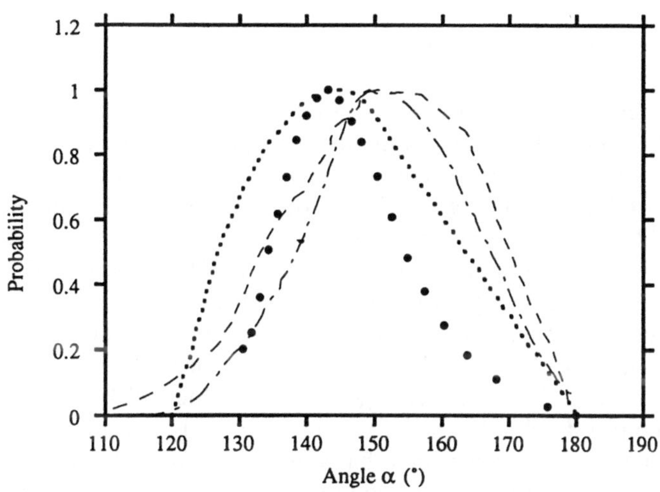

Fig. 15c. Comparison of the bond angle distribution function derived from ^{17}O DAS NMR for glassy $K_2Si_4O_9$ (*circles*) with the respective function derived from X-ray data on glassy SiO_2 (*dots*) and various molecular dynamics simulations (*dashed curves*). Reproduced with permission from Ref. [25]

(9). The distribution function of α derived on the basis of this 2-D correlation is shown as the filled circles in Fig. 15b. This analysis can be extended further to derive a more complete distribution function (open circles in Fig. 15b) if the entire lineshape in the ω_1 domain is transformed into the distribution of ^{17}O second-order nuclear electric quadrupolar shifts, taking advantage of the fact that the quadrupolar product $P_q = (e^2qQ/h)\cdot(1 + \eta^2/3)$ is relatively insensitive to the value of η. Characteristically, the bond angle distribution is significantly narrower than that obtained by X-ray diffraction on glassy silica [64]. It is also at variance with recent molecular dynamics simulations on compositionally related sodium silicate glasses [65] (see Fig. 15c).

The approach presented by these authors can in principle be extended to any oxide glass that shows resolved ^{17}O NMR spectra for the bridging oxygen sites in the isotropic dimension of the DAS experiment.

A potential alternative to measuring the bond angle distribution function is SATRAS. Because of the large ^{17}O nuclear electric quadrupole constant in SiO_2, observation and analysis of the first-order quadrupolar satellite sidebands in this material is a quite challenging task. Nevertheless, most recently Jäger et al. have reported preliminary ^{17}O SATRAS results [66], which confirm the previously deduced average isotropic chemical shift at 37 ppm vs H_2O. From the spectra shown the width of the first-order quadrupolar sidebands can be estimated as 70 ppm. Field dependent measurements of this width are expected to afford a more precise distribution function of isotropic chemical shifts. Comparison with ab-initio calculations of ^{17}O chemical shifts as a function of the Si—O—Si bond angle will then make it possible to derive a bond angle distribution function for glassy SiO_2.

3.3 *Unusual Silicon Coordination States*

Extensive in-situ high-temperature and high-pressure NMR studies have been carried out on $K_2Si_4O_9$ glass and related systems, to explore the structure of silicate melts under geologically relevant conditions. As the temperature of glassy $K_2Si_4O_9$ is raised above the glass transition, the distinct lineshape components attributable to distinct $Si^{(n)}$ species coalesce about 200 °C above T_g due to $Si^{(3)} \Leftrightarrow Si^{(4)}$ chemical exchange occurring on the NMR timescale [67]. Slower exchange rates are conveniently probed by 2-D chemical exchange NMR. Figure 16 shows a typical 2-D exchange spectrum of this material at 571 °C (close to T_g). The cross-peaks C and D shown in this plot clearly reveal that the $Si^{(4)}$ and $Si^{(3)}$ species are in chemical exchange with each other [68]. Because of the availability of empty *d*-orbitals on the silicon atom, it is reasonable to speculate that the site interchange mechanism is based on an addition/elimination mechanism, involving a higher-coordinate silicon intermediate. Support for this idea comes from the observation (Fig. 17) that the ^{29}Si NMR spectrum of $K_2Si_4O_9$ glass contains a weak peak (comprising ca. 0.06–0.10% of the total ^{29}Si signal) around −150 ppm. [69]. This chemical shift appears approximately

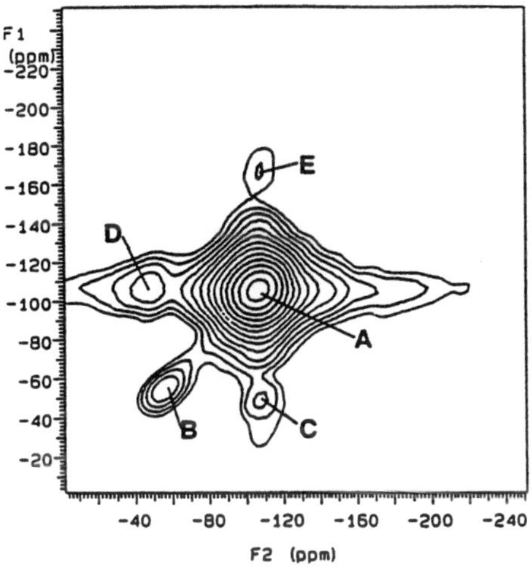

Fig. 16. Contour plot of the 2-D chemical exchange ^{29}Si NMR spectrum of $K_2Si_4O_9$ glass at 571 °C. The cross-peaks *C* and *D* reveal that the $Si^{(3)}$ and $Si^{(4)}$ species are in chemical exchange. Feature *E* possibly reflects a five-coordinated species in equilibrium with $Si^{(4)}$. Reproduced with permission from Ref. [68]

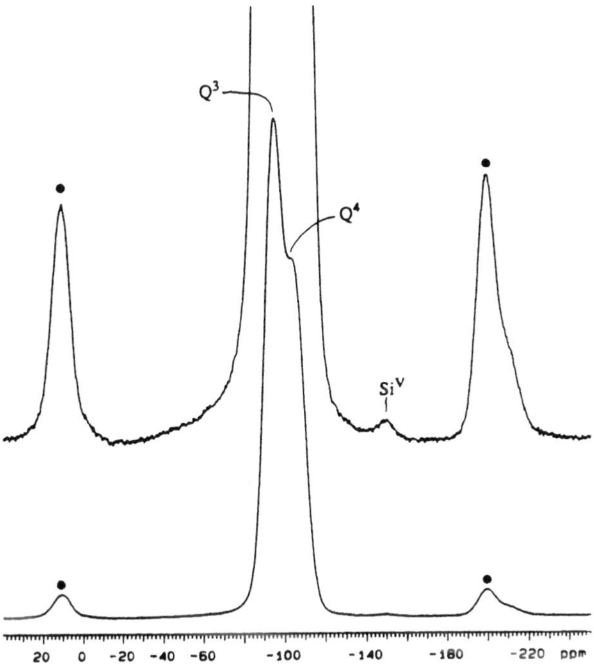

Fig. 17. ^{29}Si MAS NMR spectrum of $K_2Si_4O_9$ glass quenched at ambient pressure. The presence of the Si^V peak is indicated. Spinning sidebands are designated by dots

halfway between those measured for four- and six-coordinate silicon sites and is therefore assigned to five-coordinate silicon atoms (Si^V). At higher quenching rates, this extra peak is more intense. Substantial amounts of Si^V and Si^{VI} are present in analogous $Na_2Si_4O_9$ glasses quenched from elevated pressures (>6 GPa) hence further supporting this assigment [70]. The formation of Si^{VI}

in the GPa pressure range is also predicted by molecular dynamics simulations [71]. ^{29}Si NMR studies indicate further that the presence of the network former P_2O_5 promotes the formation of six-coordinate silicon atoms, even at ambient atmospheric pressure [72].

4 Boron Oxide-Based Glasses

Boron oxide and borate glasses are believed to be important exceptions to the continuous random network model, being characterized by substantial intermediate range order, rather than statistical site distributions. Structural studies suggest that borate glasses constitute disordered arrangements of polynuclear molecular units, known to exist in crystalline model compounds [73].

Bray and coworkers have confirmed these hypotheses using static ^{11}B and ^{10}B NMR on boron oxide and many binary and ternary borate glass systems for several decades [74]. Notwithstanding the progress in pulsed high-field NMR techniques, low-field continuous wave NMR has been most effective in

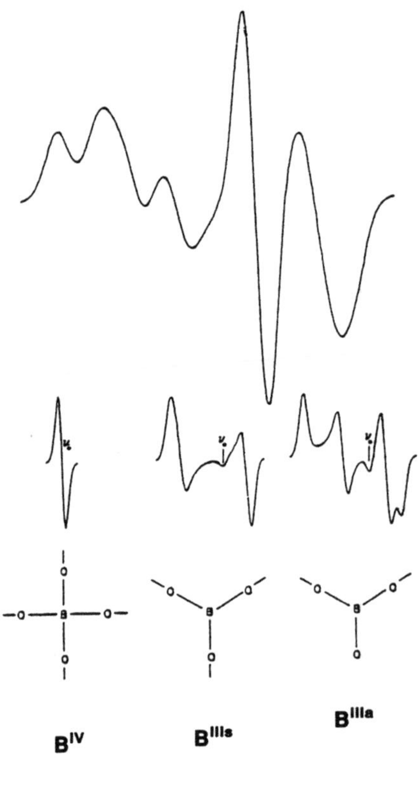

Fig. 18. Boron sites in borate glasses and their derivative ^{11}B NMR spectra as detected by cw-wideline NMR

differentiating quantitatively between three different types of boron environments on the basis of the magnitude and asymmetry of the electric field gradient at the boron sites (see Fig. 18): Symmetric-trigonal $BO_{3/2}$ groups are characterized by e^2qQ/h around 2.6 MHz and asymmetry parameter near zero. Asymmetric-trigonal $BO_{2/2}O^-$ or $BO_{1/2}(O^-)_2$ groups have e^2qQ around 2.6 MHz and $\eta \sim 0.6$, while tetrahedral $BO_{4/2^-}$ groups have generally $e^2qQ < 1$ MHz, and η is not well known. At low fields, both types of three-coordinate boron sites show well-defined powder patterns due to second-order quadrupolar perturbations, whereas four-coordinated $BO_{4/2^-}$ groups yield sharp lines with little evidence of quadrupolar broadening [74]. It is impossible here to do justice to the early pioneering NMR studies undertaken with individual borate glass systems over the past 35 years. Rather, this section will discuss insights obtained from the recent application of new experimental approaches to glassy B_2O_3 and alkali borate glasses.

4.1 Site Distribution in Glassy Boron Oxide

^{11}B, ^{10}B, and ^{17}O NMR spectra of glassy B_2O_3 have been studied extensively to investigate the extent to which six-membered boroxol rings are present. Accurate simulation of the ^{11}B and ^{10}B resonances in glassy B_2O_3 requires inclusion of a non-zero asymmetry parameter and distribution functions in both e^2qQ/h and η which are difficult to discuss independently from each other [75]. Figure 19 illustrates three novel experimental approaches as applied to glassy B_2O_3. The zero-field NQR spectrum observed with the SQUID (superconducting quantum interference device) method suggests a continuous distribution, which might obscure the discrimination between distinct boron sites (Figure 19a) [82]. In contrast, pure NQR experiments, recently developed and conducted by Bray and coworkers [76–81] clearly show two distinct boron sites with quadrupolar frequencies of 1358 and 1305 kHz, respectively (see Fig. 19b). These sites are assigned to boron atoms inside and outside boroxol rings, respectively [76]. The peak ratio suggests that 85% of all boron atoms are contained within boroxol rings. The width of the resonances shown in Fig. 19b is of considerable interest, since it reflects the distribution of quadrupolar frequencies due to topological disorder. This distribution is seen to be substantially narrower than previously inferred from the inferior method of ^{11}B NMR peak fitting [75]. Figure 19c shows a most recent result from ^{11}B DAS-NMR [83]. The sample used in this experiment was isotopically depleted in ^{11}Be to ca. 3%, in order to reduce the ^{11}B–^{11}B dipole-dipole coupling, which interfers with the experiments. This experiment detects two distinct sites in a 3:1 intensity ratio, with isotropic chemical shifts of 18.1 and 13.1 ppm, respectively. Again, the majority site is assigned to B atoms in boroxol rings, and the minority site to loose BO_3 groups. The nuclear electric quadrupole coupling constants obtained for these sites from the anisotropic dimension in the DAS experiment are in excellent agreement with the values obtained from pure NQR.

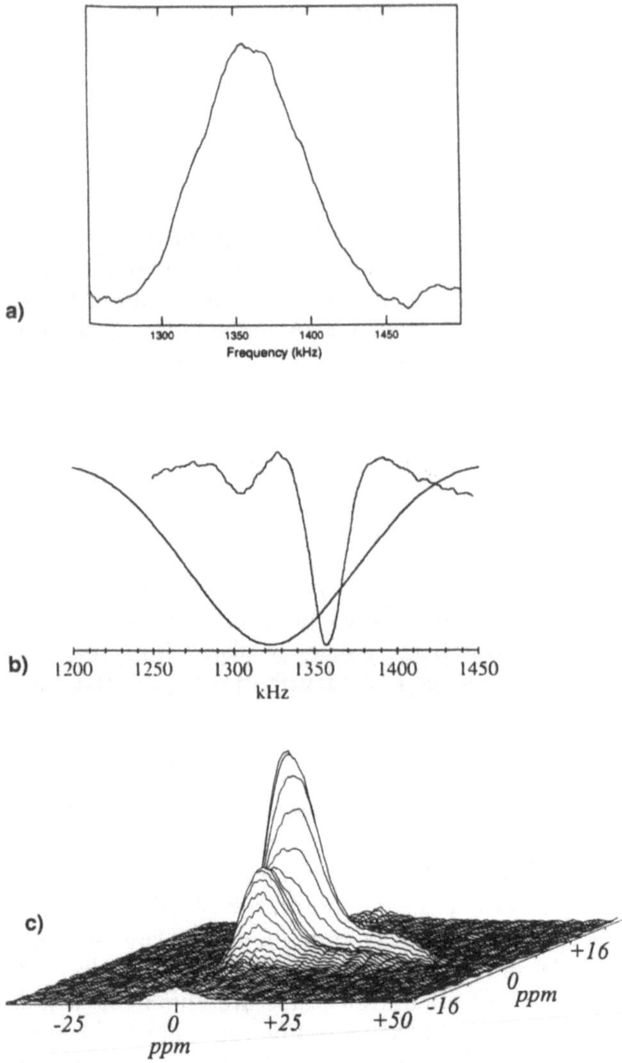

Fig. 19a–c. Three recent experimental ^{11}B NQR/NMR results relating to the structure of glassy B_2O_3: **a** SQUID detection; reproduced with permission from Ref. [82]; **b** cw-^{11}B NQR, both experimental and predicted based on previous ^{11}B NMR studies; reproduced with permission from Ref. [78]; **c** 2-D ^{11}B DAS NMR, carried out on an isotopically depleted sample; reproduced with permission from Ref. [83]. The NQR and DAS methods detect two different types of boron sites, assigned to boron atoms within boroxol rings (majority site) and loose BO_3 groups (minority site)

Additional evidence and quantitative information regarding the boroxol hypothesis might come from ^{17}O NMR. Indeed, the low-field ^{17}O NMR spectrum has been simulated assuming two different sites with quadrupole coupling constants of 4.69 and 5.75 MHz, respectively, but resolution and signal-to-noise ratios are low [75]. Ab-initio calculations have shown that the ^{17}O quadrupole coupling constant is expected to increase with increasing B—O—B angle [84, 85], and a DAS-NMR experiment as described above for silicates should offer additional insights into the fraction of boroxol rings and the distribution of B—O—B angles.

4.2 Binary Alkali Borate Glasses

The network modification of glassy B_2O_3 melts by alkaline oxides has been extremely well studied, and quantitative site ratios B^{IIIs}: B^{IIIa}: B^{IV} have been determined systematically as a function of R, the molar network modifier/network former ratio. Quite in general, in alkaline borate glasses the fraction N_4 of four-coordinated boron atoms goes through a maximum near $R = 0.5$. At higher alkaline oxide concentrations the successive growth of signal components due to B^{IIIs} units marks the increasing production of non-bridging oxygen species. While early studies noted no significant influence of the alkali ion on this speciation, more recent results, using improved lineshape deconvolution procedures, suggest that there is a specific cation effect [86]. Above $R = 0.5$, N_4 decreases systematically in the order Li, Na, K, Rb, Cs, at any given R value. In addition, mixed alkali glasses show systematically reduced N_4 values [86], indicating that B^{IIIa} units compete more successfully with B^{IV} units in the molten states.

Further structural insights into the conversion of the B_2O_3 network by alkaline oxides have recently been obtained by the cw-NQR method [78]. A typical result is shown in Fig. 20. These spectra demonstrate eloquently the successive destruction of boroxol groups by alkali oxide network modifiers, and their virtually complete elimination near $R = 0.25$. Furthermore, different types of tetrahedral B^{IV} sites can be distinguished based on their different quadrupole frequencies. Again, the resolution is much improved compared to previous wideline NMR results [74]. For different systems (crystalline and glassy), the

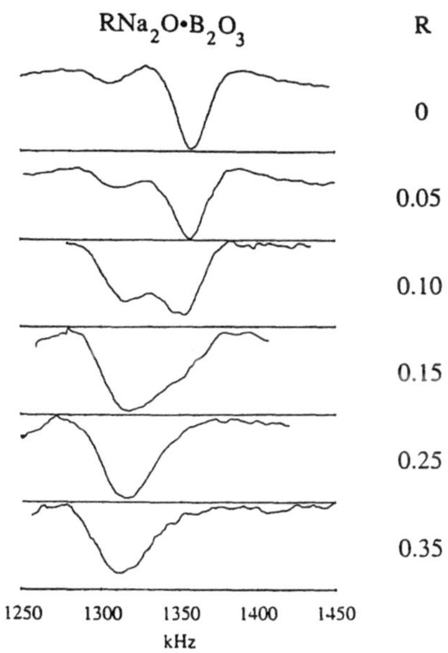

Fig. 20. ^{11}B cw-NQR spectra of sodium borate glasses as a function of R. Reproduced with permission from Ref. [78]

range of quadrupole frequencies measured for B^{IV} sites in borate glasses currently encompasses the interval $0 < \nu_Q < 687$ kHz. For individually resolved sites, it appears that the distribution functions of the quadrupole coupling parameters are substantially narrower than previously believed, suggesting the presence of some intermediate range order. Although the development of ^{11}B cw-NQR has been very recent, these preliminary studies have clearly demonstrated substantial promise for revealing further structural details of borate glasses.

Although the quantitative distinction between the various boron sites shown in Fig. 18 has been accomplished well by ^{11}B cw-NMR, this method suffers from low intrinsic sensitivity as well as from the increasing lack of availability of such spectrometers. Standard FT-wideline experiments at higher fields have generally given less satisfactory results. With increasing field strength, the width of the B^{III} resonance decreases (due to diminished second-order quadrupolar effects), while that of the B^{IV} resonance increases (due to increased chemical shift dispersion), hence making the spectral separation more difficult. Even with high-field, high speed MAS NMR one observes substantial peak overlap, because the isotropic chemical shift difference between the B^{III} and the B^{IV} units (ca. 15–20 ppm) is largely cancelled by their respective difference in $\delta^{(2)}$ which

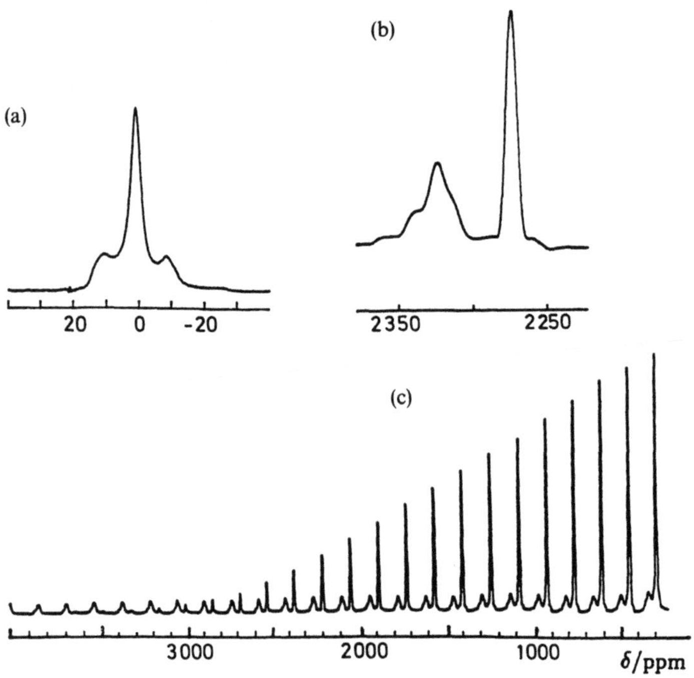

Fig. 21a–c. ^{11}B SATRAS-NMR results on a glass containing 33 mol% Li_2O and 67 mol% B_2O_3. **a** Central transition, **b** 13th sideband (downfield) of the satellite transition, **c** downfield part of the satellite transition spinning sideband spectrum. Reproduced with permission from Ref. [87]

has opposite sign. Again, this problem can be successfully addressed by SATRAS: Eq. (4) shows that the second-order shifts associated with the quadrupolar satellites are twice as large as those for the central transition, and, furthermore, have positive sign. An important consequence is that the spectral resolution between B^{III} and B^{IV} units is *enhanced* for the satellite transitions because the differences in δ_{iso} and $\delta^{(2)}$ now possess the same sign. Important new results demonstrating this resolution improvement are shown in Fig. 21 [87]. Besides quantitative reliability, the SATRAS method offers accurate chemical shift and quadrupolar coupling information on the various types of boron sites.

4.3 Ternary Borate Glasses

The presence of an additional network former or an intermediate oxide has a profound influence on the boron speciation, and such effects have been widely studied by ^{11}B cw-NMR [4]. A simple phase diagram scheme has been proposed to rationalize the experimental results [88, 89]. In essence, this scheme uses the lever rule to predict N_4 for a given glass from the experimental N_4 values of certain "standard" glasses, whose compositions correspond to those of congruently melting compounds in the ternary phase diagram. If the phase diagram is not well known, however, additional assumptions have to be introduced with regard to the composition of such "standard glasses". Whether or not such choices can always be rationalized in a plausible way, the predictive schemes developed by these authors appear to work well for a variety of ternary borate glasses and can serve as an alternative for the (less general) empirical relations used previously, particularly in the borosilicate systems [90] for describing the dependence of N_4 on composition.

5 Phosphate Glasses

The structure of phosphate glasses can be described, in a manner very similar to silicate glasses, in terms of four individual $P^{(n)}$ sites ($n \leqq 3$), varying in the number of bridging oxygen atoms. As illustrated in Fig. 22, these sites are easily distinguished by ^{31}P isotropic and anisotropic chemical shift parameters. Quantitative deconvolution of the ^{31}P MAS-NMR spectra reveal that in binary alkali and silver phosphate glasses [91, 92] and in alkaline-earth phosphate glasses [93] the $P^{(n)}$ site distribution follows the binary model. Since water acts as a network modifier in these hygroscopic glasses, a precise analysis of NMR spectra of these glasses requires the exact water content to be known. Most recently, Brow and coworkers have confirmed that the binary model is also applicable in alkali phosphate glasses prepared and stored under rigorously anhydrous

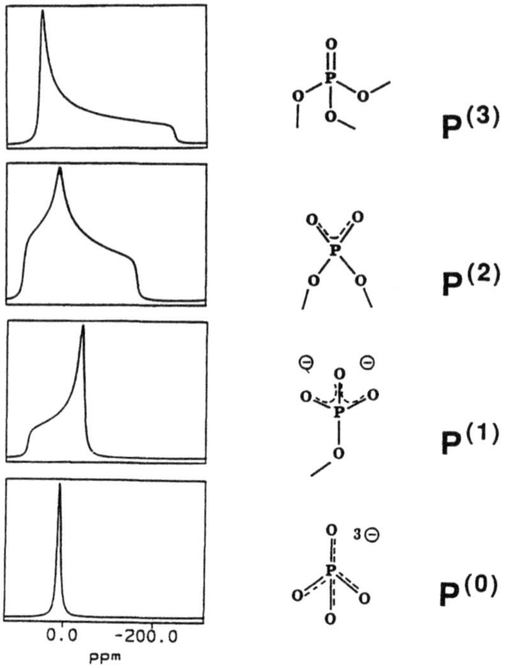

Fig. 22. Phosphorus sites in alkali phosphate glasses and their wideline ^{31}P NMR anisotropic chemical shift powder patterns. The terminology used here is analogous to that defined in Fig. 11

conditions [94]. Zwanziger and coworkers have utilized the VACSY technique to correlate specific isotropic chemical shifts with the corresponding ansisotropic chemical shift information in ionically conducting silver phosphate glasses [27]. Figure 23 shows an interesting result for a glass with composition $(Ag_2O)_{41}$ $(AgI)_{39}(P_2O_5)_{20}$. The MAS–NMR spectrum shows a low-intensity peak at 23.6 ppm, assigned to PO_4^{3-} ($P^{(0)}$) groups, and a dominant resonance centered around 3 ppm, assigned to $P_2O_7^{4-}$ ($P^{(1)}$) groups. Both peaks are broadened by a distribution of isotropic chemical shifts, reflecting the inherent disorder in the glass. It is evident from Fig. 23 that VACSY is able to recover the individual CSA information for these two distinct sites. Even more interesting, however, the experiment reveals that the chemical shift tensor elements for the $P^{(1)}$ resonance depend on the exact position within the isotropic chemical shift distribution, larger anisotropies being associated with more upfield shifts. Since it is known that the ^{31}P chemical shift anisotropy is closely correlated with the P=O bond length [95], the VACSY experiment allows one to estimate the distribution function of this parameter in phosphate glasses. Finally, the most upfield wing of the $P^{(1)}$ resonance is correlated with a non-axially symmetric chemical shift tensor, revealing the presence of distinct, more distorted pyrophosphate groups. The demonstrated ability of this technique to enhance the resolution between sites whose MAS-NMR centerbands overlap, makes it a potentially very powerful tool for the investigation of many other spin-1/2 nuclei in glasses and other disordered materials.

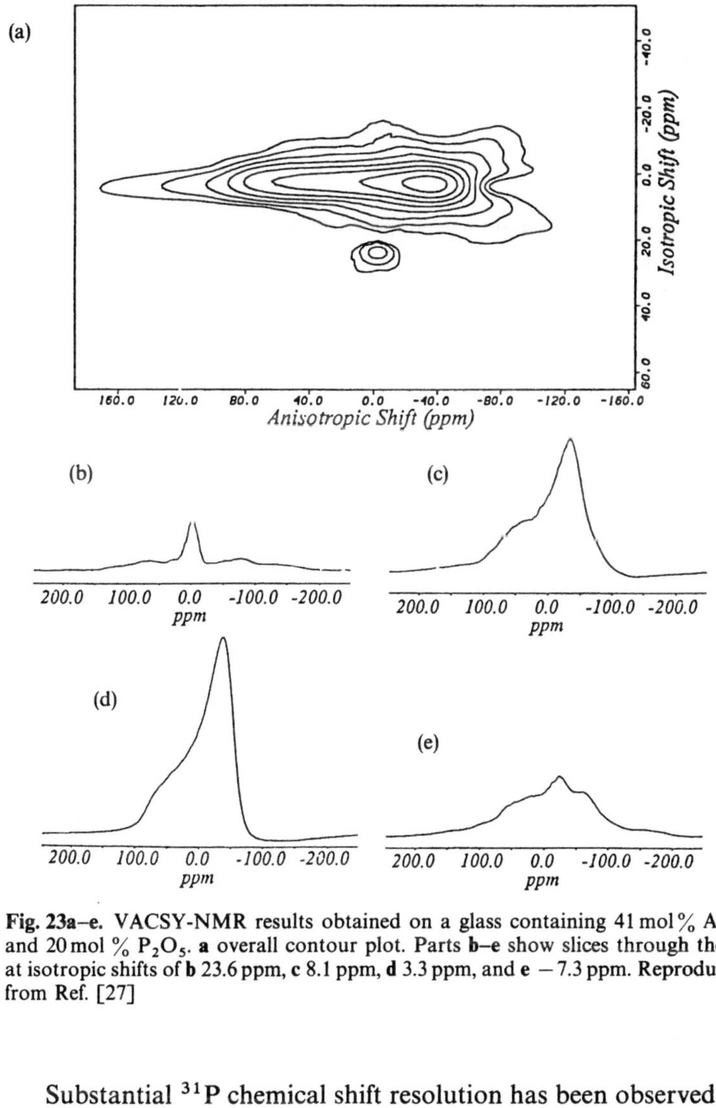

Fig. 23a–e. VACSY-NMR results obtained on a glass containing 41 mol% Ag_2O, 39 mol% AgI, and 20 mol % P_2O_5. **a** overall contour plot. Parts **b–e** show slices through the VACSY spectrum at isotropic shifts of **b** 23.6 ppm, **c** 8.1 ppm, **d** 3.3 ppm, and **e** −7.3 ppm. Reproduced with permission from Ref. [27]

Substantial ^{31}P chemical shift resolution has been observed in boro-and (to a lesser extent) aluminophosphate glasses. Considering next nearest neighbor environments, these glasses may contain (depending on composition) up to eleven sites $P_m^{(n)}$ where m is the number of boron or aluminum next nearest neighbors ($m \leq n$). Among these sites, a feature not known for binary alkali phosphate glasses are the tetrahedral, fully bridging $P_4^{(4)}$ sites, well known in crystalline BPO_4 and $AlPO_4$. The ^{31}P MAS-NMR spectra of silver borophosphate glasses were found to be exceptionally well-resolved [96], permitting a quantitative distinction between various of these sites. In contrast, only partial site resolution is observed for sodium aluminophosphate glasses [97, 98]. It can be anticipated that applications of VACSY and ^{27}Al-^{31}P double resonance techniques will be particularly powerful in arriving at site assignments and quantifications in future studies.

Finally, a number of multinuclear NMR studies have been reported recently on more complex phosphate glass systems. Contrary to the behavior of ^{29}Si in alkali silicate glasses, the ^{31}P chemical shift in phosphate glasses depends strongly on the nature of the counter-cation [4, 99]. The high sensitivity of ^{31}P chemical shift to the chemical environment has been exploited further in the study of various anion-substituted glass systems, most notably the technologically important fluorophosphate glasses [100–102]. So far, the work on the latter systems has remained confined to simple MAS-NMR studies, and the current interpretations must be regarded as preliminary. Fluorophosphate glasses represent, however, an ideal system for exploiting the full arsenal of dipolar and double resonance techniques solid state NMR has to offer, and substantial progress in this area is anticipated.

Aluminum in Oxide Glasses

6.1 Bond Ordering in Aluminosilicate Glasses

Aluminum oxide, while unable to form glass by itself via melt quenching, is an important constituent of many alkali and alkaline-earth silicate glasses of geological and practical interest. An abundance of NMR studies on these glasses have been published, and it is not possible here to enumerate all the details observed for individual systems [4]. Rather, the focus here will be on the most recent results and the broader chemical conclusions. From a structural point of view, it is practical to subdivide any system $M_{(2)}O—Al_2O_3—SiO_2$ (M = alkali or alkaline earth) into four compositional regions (shown in Fig. 24):

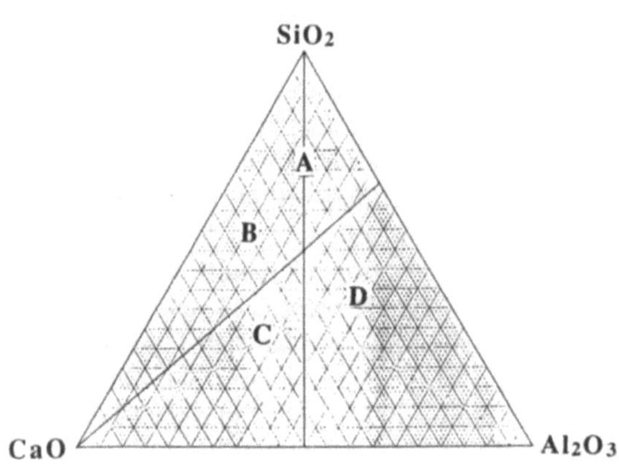

Fig. 24. Glassforming region in the SiO_2–CaO–Al_2O_3 system. The four compositional regions A–D discussed separately in this section are indicated

Region A: $Al_2O_3/SiO_2 \leqq 0.5$ and $M_{(2)}O/Al_2O_3 = 1$.
Region B: $Al_2O_3/SiO_2 \leqq 0.5$ and $M_{(2)}O/Al_2O_3 > 1$.
Region C: $Al_2O_3/SiO_2 > 0.5$ and $M_{(2)}O/Al_2O_3 \geqq 1$.
Region D: $M_{(2)}O/Al_2O_3 < 1$.

6.1.1 Region A: Fully Polymerized Tetrahedral Networks

Most ^{29}Si and ^{27}Al MAS-NMR studies have supported the pre-existing view that such glasses are fully polymerized networks comprised of $Si^{(4)}$ units and $MAlO_2$ fragments ($AlO_{4/2}^{-}$ tetrahedra whose negative charge is balanced by cations) [103–108]. The ^{29}Si MAS-NMR spectra of such glasses are poorly resolved and the center of gravity is shifted downfield with increasing $MAlO_2$ content, from -110 to -86 ppm with increasing Al_2O_3/SiO_2 ratio, reflecting the successive replacement of Si—O—Si bonds by Si—O—Al linkages. Similar chemical shift trends have been observed experimentally in zeolites. All of these results are consistent with the Lowenstein rule [109], according to which Al^{IV}—O—Al^{IV} linkages in the framework, which represent an energetically unfavorable accumulation of charge in the network, are being avoided.

Recently, unusual behavior was found in slightly peraluminous rapidly quenched samples along the $MgAl_2O_4$—SiO_2 join, indicating substantial amounts of five and six-coordinated aluminum atoms [110].

6.1.2 Region B: Chemically Modified Tetrahedral Networks

These glass compositions have excess modifyer oxide, which in principle could react with either the SiO_2 or the $MAlO_2$ fragments. This question has been investigated extensively, but earlier MAS-NMR studies have generally suffered from assignment ambiguities, because both Al substitution and formation of non-bridging oxygen atoms result in downfield chemical shift effects of comparable magnitude. Figure 25 shows some recent wideline NMR studies of the system $4SiO_2$—$(Na_2O)_{1-x}(Al_2O_3)_x$ ($0 \leqq x \leqq 0.5$) [111]. As the Al/Na ratio is decreased from 0.5 to 0.1, the axially symmetric chemical shift pattern of $Si^{(3)}$ units emerges with its quantitatively expected contribution. Thus the excess Na_2O in this system associates primarily with the SiO_2 fragments. This behavior appears to be quite general. In Fig. 26 the number of non-bridging oxygen atoms per tetrahedral atom (obtained by NMR peak area analysis) is plotted as a function of $N_{Al} = 4[Al]/\{[Al] + [Si]\}$. These data are compared with (a) a model assuming that the excess sodium oxide modifies the Si—O—Si bonds exclusively (solid lines) and (b) a model assuming random association of Na_2O with both Si and Al atoms. Clearly the data are much more consistent with the first assumption and support a model where variable $Si^{(n)}$ units are interlinked with fully polymerized AlO_4^{-} tetrahedra.

The overall ^{29}Si chemical shifts in these glasses are influenced by the effects of both nonbridging oxygen atoms and next-nearest Al atoms. According to

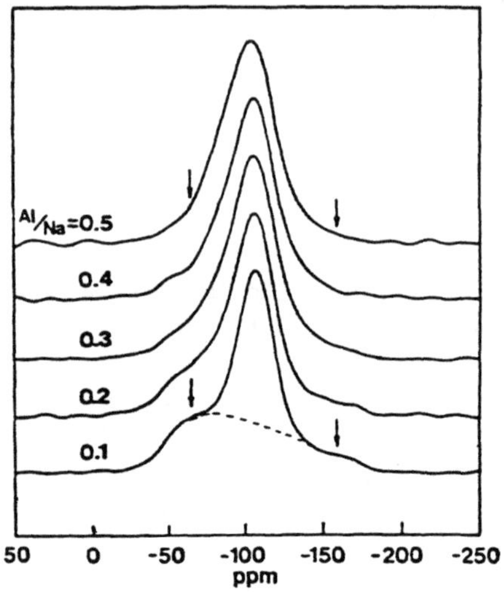

Fig. 25. Wideline ^{29}Si NMR in aluminosilicate glasses along the compositional series $(1-x)$ $Na_2O \cdot xAl_2O_3 \cdot 4SiO_2$. Note the emergence of a $Si^{(3)}$ powder pattern (chemical shift principal components indicated by *arrows*) at low Al/Na ratios. Reproduced with permission from Ref. [111]

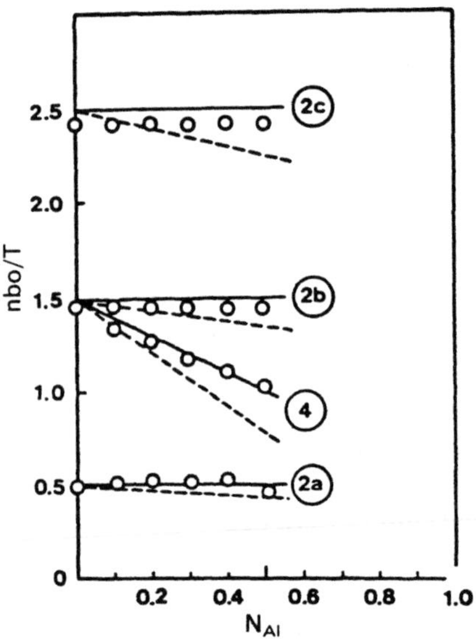

Fig. 26. Numbers of non-bridging oxygen atoms per tetrahedral atom (Al + Si) obtained from NMR peak area analysis (*circles*) in sodium aluminosilicate glasses. *Solid line*: calculated number assuming that non-bridging O atoms are exclusively associated with Si atoms. *Dashed line*: calculated number assuming that non-bridging O atoms are associated with both silicon and aluminum atoms. Reproduced with permission from Ref. [111]

Maekawa et al., these effects are approximately additive, and can be expressed numerically by a series of coefficients:

$$\delta = N_{nbo}\partial_{Na} + N_{Al}\partial_{NaAl} + C_0 \tag{10}$$

where $C_0 = -110.0$ ppm (chemical shift of pure SiO_2 glass), ∂_{Na} and ∂_{NaAl} are chemical shift increments induced by Na^+ (nbo formation) and $NaAlO_2$ neighbors (Al-O-Si formation). Both ∂_{Na} and ∂_{nbo} depend further on the glass

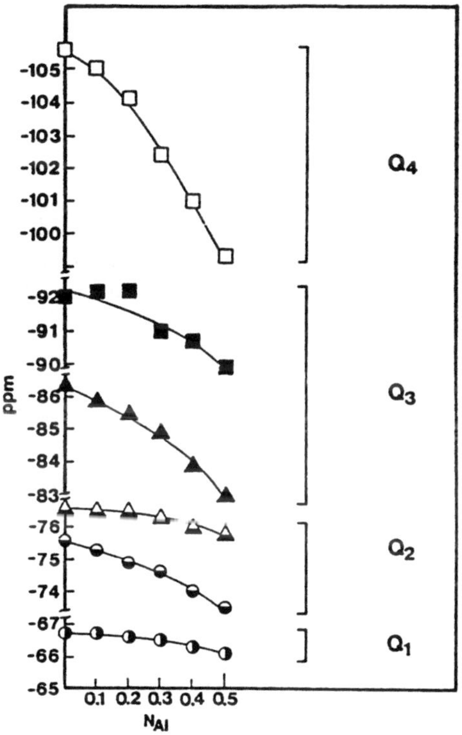

Fig. 27. ^{29}Si chemical shifts of the $Q^{(n)}$ (or $Si^{(n)}$) units as function of $N_{Al} = 4\ Al/(Al + Si)$ for a series of sodium aluminosilicate glasses. Reproduced with permission from Ref. [111]

composition:

$$\partial_{Na} = C_1 N_{nbo} + C_2 N_{Al} + C_3 \quad (11a)$$

$$\partial_{NaAl} = C_2 N_{nbo} + C_4 N_{Al} + C_5 \quad (11b)$$

All of the coefficients C_0–C_5 have been determined by least-squares fitting the weighted mean ^{29}Si chemical shifts of 54 glasses [111]. This description allows approximate chemical shift predictions in aluminosilicate glasses [111]. Simpler, albeit less accurate chemical shift analyses have been attempted [110].

Finally, Fig. 27 suggests that when $NaAlO_2$ is introduced into a glass with a given $Si^{(n)}$ distribution, the chemical shifts of the more polymerized species (higher n) are much more strongly affected than those of the less polymerized species (lower n). Although a variety of explanations are possible for this observation, Maekada et al. believe that it reflects a preference of Al to associate with more polymerized silicon sites [111]. More conclusive proof for this assertion will have to await detailed Si-Al double resonance NMR experiments.

6.1.3 Region C: Alumina-Rich Charge-Balanced Aluminosilicate Glasses

The region of glass formation in the system CaO–Al_2O_3–SiO_2 extends all the way to glassy $CaAl_2O_4$. This material is of interest for long wavelength optical fiber applications. The Al atoms are tetrahedrally coordinated [112, 113] as found in crystalline $CaAl_2O_4$. The corresponding magnesium compound has six-coordinated Al atoms and it is impossible to vitrify it. Ultrahigh temperature NMR studies show, however, a large chemical shift difference between both liquids, possibly related to differences in the average coordination numbers (see Fig. 28) [114]. Chemical shift trends along the $CaO{-}Al_2O_3$ tie line have been interpreted in the same vein [113].

To reduce the innate tendency of $CaAl_2O_4$ to devitrify, SiO_2 is added, resulting in peculiar effects on the glass transition temperature, with a maximum observed around 10–15 mol% SiO_2 [115]. The ^{29}Si chemical shifts in the silica-poor region lie between −75 and −78 ppm, consistent with either $Si^{(4)}(OAl)_{4/2}$ units or with depolymerized $Si^{(2)}$ groups. The static spectrum shows no sign of a chemical shift anisotropy expected for a $Si^{(2)}$ site [116]. This would support the former assignment, however, the authors caution that the apparent lack of chemical shift anisotropy in the spectrum could be caused by disorder effects. In such a situation, the separation of isotropic and anisotropic chemical shift components by the two-dimensional VACSY technique might be the

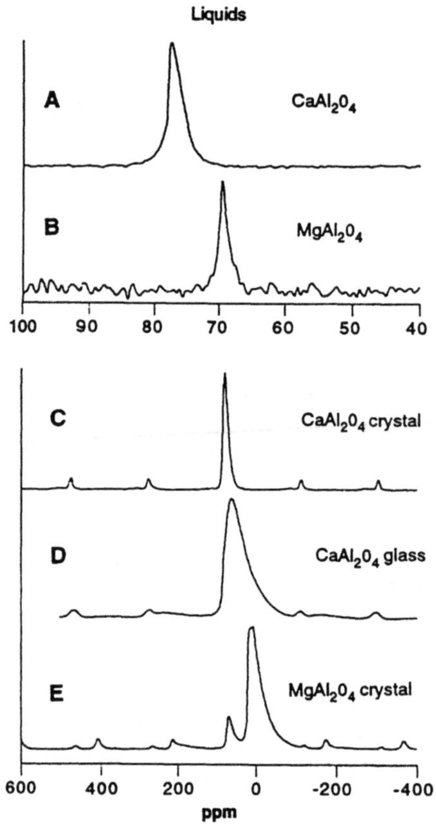

Fig. 28. Ultra-high temperature (2500 K) spectra of *A* liquid $CaAl_2O_4$, *B* liquid $MgAl_2O_4$. *C-E*: Solid state ^{27}Al MAS-NMR spectra of *C* crystalline $CaAl_2O_4$, *D* glassy $CaAl_2O_4$, and *E* crystalline $MgAl_2O_4$. Reproduced with permission from Ref. [114]

technique of choice to confirm this conclusion. If part of the CaO were indeed used to generate non-bridging oxygen atoms associated with $Si^{(4-n)}$ groups $(n > 0)$, not enough CaO would be available to charge-balance the AlO_4^- groups, and thus Al in higher coordination numbers should be observed. Results previously published for glasses in this compositional region, particularly for $CaAl_2SiO_6$ glass, are contradictory in this regard [103, 104].

McMillan and coworkers observe that the ^{27}Al MAS-NMR lineshapes of aluminosilicate glasses in region C show a characteristic spinning speed dependence, with the apparent resolution decreasing as the spinning frequency is increased (see Fig. 29). [110] This behavior can be understood on the basis of a wide distribution of nuclear quadrupole coupling constants. At low spinning speeds, only the most symmetric Al sites with small quadrupole couplings experience narrowing, whereas the others contribute only to a broad unresolved hump above the baseline. As the spinning frequency is increased, however, more and more Al sites with stronger quadrupole couplings contribute to the narrowed MAS centerband. As these signals begin to contribute to the MAS centerband at higher spinning speeds, the overall width of the MAS centerband increases, because for such larger quadrupole coupling constants, MAS can narrow the second-order quadrupolar powder patterns only partially. These results illustrate again the necessity of applying the fastest possible spinning speeds and the strongest magnetic fields for ^{27}Al MAS-NMR studies of glasses.

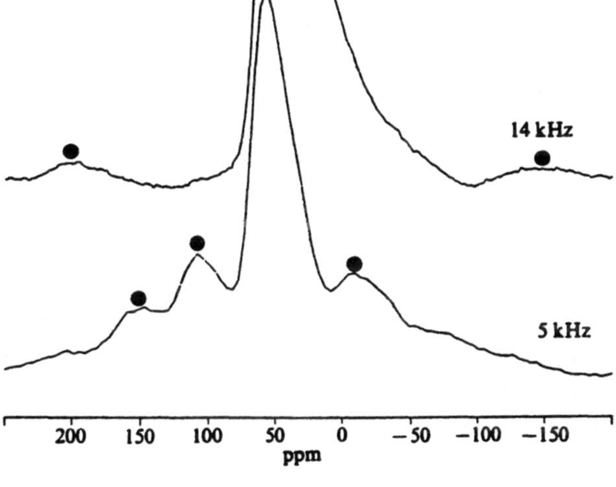

Fig. 29. Effect of spinning speed on the ^{27}Al NMR spectrum of a glass containing 15 mol% CaO, 35 mol% Al_2O_3, and 50 mol% SiO_2. Spinning sidebands are indicated by *circles*. Reproduced with permission from Ref. [110]

6.1.4 *Region D: Peraluminous Aluminosilicate Glasses*

Aluminosilicate glasses in which the charge associated with an AlO_4^- tetrahedral unit cannot be balanced (M/Al ratio below one) are of special structural interest. The system can respond to this situation either by forming segregated phases, or by altering the structural role of the aluminum atoms. For example a disproportionation of the kind

$$2[Al_2O_3] \rightarrow 3[AlO_{4/2}]^- + Al^{3+} \tag{12}$$

can be invoked, where the Al^{3+} formed is assumed to be six-coordinated. For charge-balance reasons, these reaction products might associate with each other, resulting in charge-neutral $Al(AlO_2)_3$ "triclusters" as postulated by Lacy [117]. Indeed, previous ^{27}Al NMR studies in this compositional region have shown that part of the aluminum appears with coordination numbers 5 and 6 in all peraluminous glasses. These higher coordinated Al atoms give rise to resonances near 30 and 0 ppm, respectively. In general, it appears that substantial amounts of five-coordinate Al sites are present, whereas clear evidence for Al(6) is only observed at very low M contents, near the Al_2O_3–SiO_2 tie line.

6.1.5 *Binary SiO_2–Al_2O_3 Glasses*

Among the peraluminous aluminosilicate glasses, the binary glasses formed in the system $(SiO_2)_{1-x}$–$(Al_2O_3)_x$ deserve special discussion. These glasses are of commercial importance in preparing refractory ceramics and also represent a prototype system for understanding more complex aluminosilicates. A region of metastable liquid-liquid immiscibility exists within the region $0.1 \leqq x \leqq 0.5$, resulting in glasses whose structure and composition may be dependent on the quenching rate [118–120]. This expectation is clearly borne out in the spectra shown in Fig. 30. [121]. The "superquenched" (SQ) glasses, which have experienced cooling rates of 10^5 °C/s, appear to be more (although not completely) homogeneous than the normally quenched glasses (NQ, 10^2 °C/s) [121]. These glasses show aluminum in four-, five, and sixfold coordination, whereas for the phase-separated glasses the NMR spectra reveal primarily Al^{IV} and Al^{VI}. Spin counting studies and comparison with reference standards reveal further that in the SQ samples only ca. 70–85% aluminum atoms are observable, whereas in the NQ samples the detection is quantitative [121]. There are also large differences in the ^{29}Si NMR spectra: in the NQ samples the Si environments resemble those in glassy SiO_2, whereas in the SQ samples downfield shifted signal contribution indicate either non-bridging oxygen atoms or Si—O—Al connectivities [121]. Overall, the striking influence of the quenching rate reveals that this system offers different pockets of stability in configuration space. Apparently the ordering that accompanies the unmixing of the two liquid phases disfavors the presence of Al^V and possibly other highly distorted sites. For very low Al_2O_3 contents, where complications arising from liquid-liquid immiscibility are absent, experi-

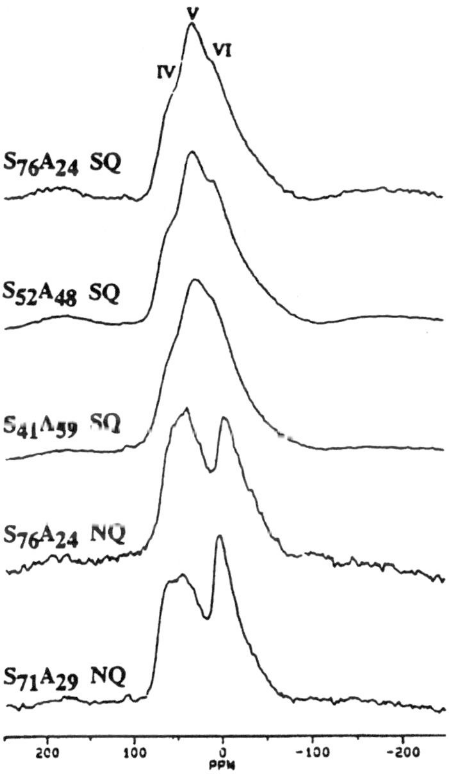

Fig. 30. ^{27}Al MAS NMR on various SiO_2–Al_2O_3 glasses. The symbols SQ and NQ denote "superquenched" and "normally quenched" glasses. mol % SiO_2(S) and Al_2O_3 (A) are given for each spectrum. Peak assignments to various Al coordination polyhedra are shown. Reproduced with permission from Ref. [121]

mental evidences are contradictory to date: Some authors observe both Al^V and Al^{VI} in accordance with {12}, but in different proportions [120, 122], whereas others detect primarily Al^{IV} [123].

The prevalence of Al^{IV} at these low Al concentrations is qualitatively in accordance with molecular dynamics simulations, which predict that the average Al coordination number in $(SiO_2)_{1-x}(Al_2O_3)_x$ liquids should increase from below 4 to ca. 5 as x increases from zero to one [124]. This prediction, which is borne out qualitatively by the MAS-NMR data of the homogeneous and SQ samples [123], has recently been tested by ultrahigh temperature ^{27}Al liquid state NMR [124]. In the liquid state, all sites exchange rapidly and thus only an average resonance is observed. Surprisingly, this resonance is found to shift slightly downfield with increasing x, contrary to the expectation from the MD simulations.

The limiting composition $x = 1$ (pure amorphous Al_2O_3) cannot be prepared from the molten state, but is accessible either by electrochemical precipitation [125, 126] or sol-gel chemistry [127, 128]. Again, ^{27}Al NMR studies reveal aluminum to be present in four-, five-, and six-coordination in such materials.

To date, the detailed structural roles of these aluminum coordination states in amorphous Al_2O_3, SiO_2–Al_2O_3 glasses, and ternary aluminosilicate glasses are altogether not well-understood and more sophisticated experiments require application. For example, Al^{IV}-Al^{VI} coherence transfer experiments might be used to test the tricluster concept, and selective $^{27}Al^{IV} \rightarrow {}^{29}Si$, $^{27}Al^{V} \rightarrow {}^{29}Si$, and $^{27}Al^{VI}$–^{29}Si coherence transfer and heteronuclear correlation experiments might be able to clarify the structural relationship between these aluminum sites and the Si framework atoms. Furthermore, charge balance considerations make it highly likely that these glasses contain three-coordinated oxygen atoms, particularly near Al^{V} and Al^{VI} sites. It is expected that future ^{17}O DAS- or DOR-NMR and ^{27}Al–^{17}O double resonance experiments will make it possible to quantify such environments.

6.2 Other Aluminosilicate Glass Systems

6.2.1 Fluorine-Containing Aluminosilicate Glasses

These are important model systems for geological glasses, since fluorine is an abundant volatile component of natural magmas. Similarly to water, the presence of fluorine reduces the melt viscosity and increases the diffusion coefficients, even at small concentrations. The location of the fluorine species in

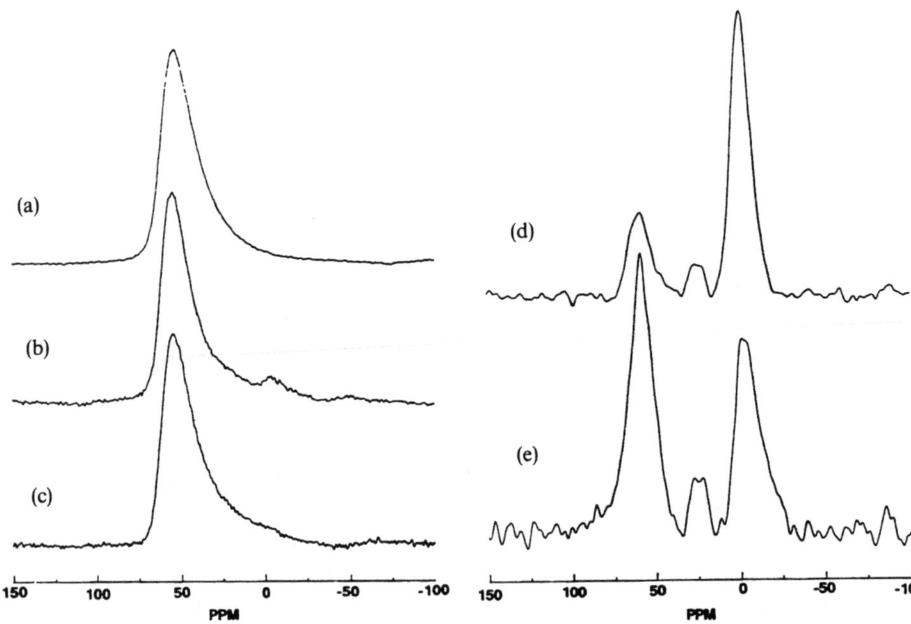

Fig. 31a–e. ^{27}Al MAS-NMR results on F-bearing jadeite glasses. *Left side*: single-pulse experiments **a** F-free jadeite glass, **b** jadeite glass, containing NaF, **c** jadeite glass, containing cryolite. *Right side*: ^{19}F-^{27}Al CPMAS-NMR experiments: **d** jadeite glass containing NaF, **e** jadeite glass containing cryolite. Reproduced with permission from Ref. [129]

geologically relevant aluminosilicate glasses has recently been addressed by multinuclear MAS-NMR techniques, including $^{19}F \rightarrow {}^{27}Al$ and $^{19}F \rightarrow {}^{29}Si$ cross-polarization experiments. Figure 31 shows interesting ^{27}Al NMR results for hydrous jadeite-NaF and jadeite-cryolite glasses. Cross-polarization from ^{19}F dramatically enhances the spectral contribution from five- and six-coordinated Al atoms, hence suggesting that fluorine preferentially coordinates with Al atoms. In general agreement with these conclusions, $^{19}F \rightarrow {}^{29}Si$ CPMAS experiments on fluorine-containing albite glasses show no evidence for the formation of Si-F bonds; species containing such connectivities would be expected to have much larger cross-relaxation rates than are actually observed in the experiments [130]. Both groups speculate that the effect of fluorine on the melt viscosity is due to a depolymerization process resulting in the formation of AlF_5^{2-} and AlF_6^{3-} species. The exact nature of the species formed has been discussed recently on the basis of ab-initio calculations of the ^{27}Al nuclear electric quadrupolar coupling constants [131].

6.2.2 *Rare-Earth Aluminosilicate Glasses*

These glasses possess unique mechanical strengths, high glass transition temperatures and great chemical durability [132]. To explore whether these properties are related to specific structural features, ^{27}Al NMR spectra have been obtained. Preliminary results seem to indicate that the rare-earth ion has a distinct influence upon the aluminum speciation. While in a glass containing 20 mole% Nd_2O_3, 20 mol% Al_2O_3 and 60 mol% SiO_2 only Al^{IV} sites are detected [133], analogous yttrium aluminosilicate glasses show Al^{IV}, Al^{V} and Al^{VI} for most compositions [134]. Obviously, reaction (12) can be written for both alumina and the rare earth metal ion, and the aluminum speciation depends on the relative preferences of Al^{3+} and RE^{3+} to be either four- or six-coordinated.

6.3 *Aluminoborate Glasses and Related Systems*

Glasses of composition $M_{(2)}O$–Al_2O_3–B_2O_3 (M = alkali or alkaline earth metal) show consistently high concentrations of Al^{V} and Al^{VI}, even if the $M_{(2)}O/Al_2O_3$ ratio exceeds unity. This is an important difference to the situation in aluminosilicate glasses, where the formation of AlO_4^- takes precedence over any other network conversion process. For aluminoborate glasses, both early ^{11}B wideline [135] and more recent ^{27}Al MAS [136] NMR studies indicate that the excess oxide introduced with M_2O or MO modifiers into the glass network is utilized to generate the formation of both AlO_4^- (Al^{IV}) units and BO_4^- (B^{IV}) units. These conversion processes are competitive, however, and a striking dependence on the counter-cation is observed [136e].

Recently, a new, detailed NMR study of alkaline earth aluminoborates has appeared [137]. In interpreting their NMR results, Brow and coworkers estimate the stability of various oxygen nearest neighbor environments on the basis of bond valence considerations [138]. Maximum stability is expected if the bond valence sum (charge divided by coordination number) of the cations next to an oxygen site are close to +2 for a neutral oxygen site, and +1 for a non-bridging oxygen site. Such considerations show, for instance, that B^{IV}–O–B^{IV} and Al^{IV}–O–Al^{IV} connectivities are less stable than B^{III}–O–Al^{IV} linkages. Furthermore, three-coordinate oxygen atoms bonded to $2B^{IV}$, $1Al^{V,VI}$ or $1B^{III}$, $2Al^{V,VI}$ have high intrinsic stability. Basically, the presence of three-coordinate oxygen atoms accounts satisfactorily for the observation of Al^{V} and Al^{VI} in this system (and many others). Still, a moderately large number of stable configurations remain, and this probably accounts for the large counter-cation effects observed in this system (see, for instance Fig. 32).

Bond valence considerations can also be used to explain some of the observed chemical shifts: As shown in Fig. 33 the chemical shift for the Al^{IV} site shows an excellent correlation with the bond valence sum for the next nearest neighbor cations. Based on such considerations, the shifts around 55–60 ppm observed for Al^{IV} in many these glasses are much more consistent with Al^{IV}-O-B^{III} linkages than with the Al^{IV}–O–B^{IV}, Al^{IV}–O–Al^{IV}, and Al^{IV}–O–Al^{VI} linkages previously proposed. These conclusions are also in agreement with the relative stability arguments outlined above. The relative population ratio of the various aluminum coordination sites is shown to depend on (1) the nature of the network cation, (2) the Al fraction in the total network former, and (3) the amount of network modifier. Further studies using spin counting will be necessary in order to address the previously-raised question of how much of the total aluminum content is actually observable. To develop a comprehensive structural model

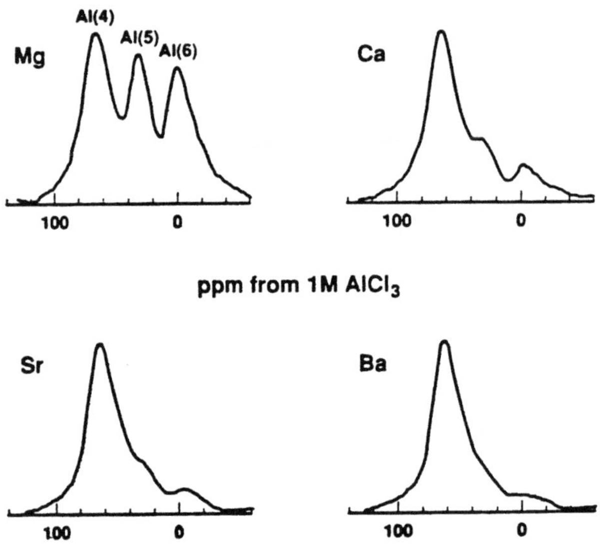

Fig. 32. ^{27}Al MAS-NMR spectra for alkaline-earth aluminoborate glasses with composition 40 mol% MO, 40 mol% B_2O_3, 20 mol% Al_2O_3, indicating the influence of the alkaline earth metal atom M upon the Al speciation. Reproduced with permission from Ref. [137]

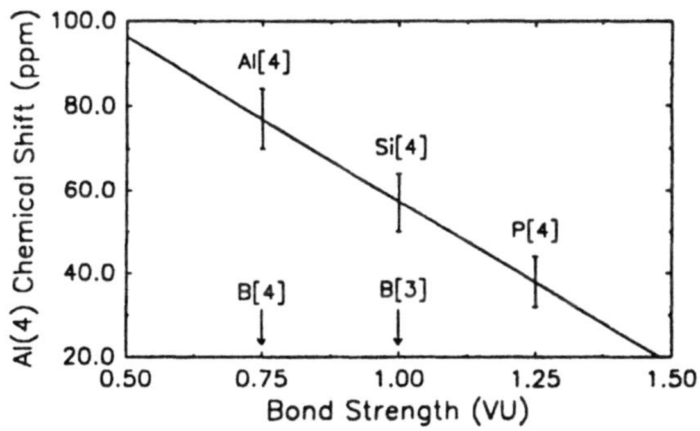

Fig. 33. Correlation of the ^{27}Al chemical shift ranges observed for Al^{IV} in framework structure crystalline model compounds (4 network next nearest neighbors) with the bond valence sum ("bond strength") of the next-nearest neighbor cation. From this plot it is predicted that an Al^{IV} species linked to four B^{IV} or to four B^{III} species should resonate at 75 ppm and 55 ppm, respectively. Reproduced with permission from Ref. [138]

for these glasses, the oxygen coordination numbers must be independently quantified, using ^{17}O DAS or DOR experiments.

The situation in sodium and calcium aluminophosphate glasses resembles that in aluminoborate glasses. Substantial fractions of Al^{V} and Al^{VI} are detected [139, 140], and their presence can be understood in terms of bond valence arguments [98].

6.4 Future Prospects of ^{27}Al NMR in Glasses

In principle, the shortcomings in ^{27}Al MAS-NMR spectroscopy with regard to more quantitative applications can be overcome to some extent by more sophisticated mechanical rotation experiments such as DAS and DOR. However, while those techniques produce substantially improved spectral resolution, they do not eliminate the second-order quadrupolar shifts (and distributions thereof) thus affecting the central transition lineshapes in glasses in an unpredictable way, making quantitative peak deconvolutions difficult. In contrast, SATRAS minimizes the second order quadrupolar effects, but does not completely eliminate the anisotropy of the second-order quadrupolar perturbations on the lineshape. Nevertheless, substantial resolution improvement has been demonstrated recently by Jäger et al. (see Fig. 34) in a variety of aluminoborate glasses [142]. Benefits are anticipated particularly in those glasses, where doubts exist, on whether all of the aluminum sites present are in fact detected in the central transition.

A second advantage of SATRAS is that it renders more accurate isotropic chemical shifts. Using this method, Lippmaa and coworkers have shown that the ^{27}Al chemical shifts of tetrahedral aluminum in framework aluminosilicates depend linearly on the average Al—O—Si bond angle θ in the following fashion

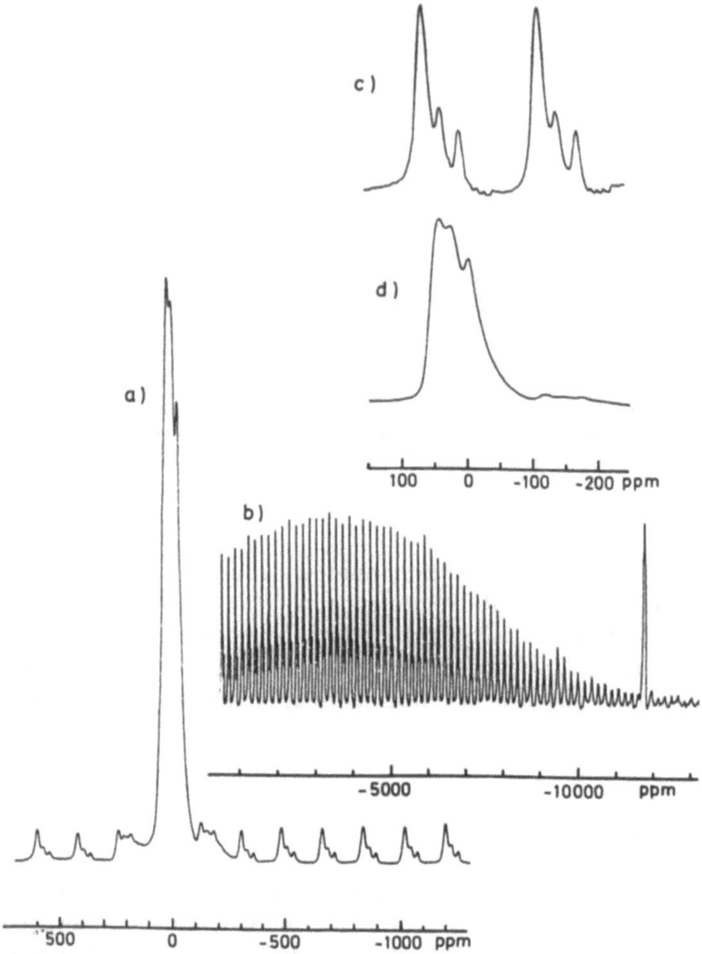

Fig. 34a–d. ^{27}Al SATRAS results on a lead aluminoborate glass with composition: 36 mol% PbO, 44 mol% B_2O_3, 20 mol% Al_2O_3. **a** Central transition and a series of spinning sidebands of the satellite transitions, **b** upfield part of the total satellite transition sideband spectrum, **c** third and fourth satellite transition sideband, **d** central transition. Note the improved resolution between Al^{IV}, Al^{V}, and Al^{VI} sites in the satellite sidebands. Reproduced with permission from Ref. [142]

[38]:

$$\delta = 132\,\text{ppm} - 0.50\,\theta \tag{13}$$

Thus, application of SATRAS to glasses allows one to measure the distribution function of isotropic chemical shifts in a selective fashion and correlation (13) can be used to relate this distribution function to a distribution function of inter-tetrahedral bond angles.

Based on these new exciting developments a considerable amount of new activity in ^{27}Al NMR of glasses can be foreseen. With the increasing availability of high-speed MAS NMR and the ability of tightly controlling spinning speeds over long measurement times it can be predicted that within a few years ^{27}Al central transition MAS NMR will have been but entirely displaced by SATRAS and, possibly, DOR. We will see much of the central-transition MAS-NMR work surveyed in this section repeated with these techniques, substantially improving our understanding of the structural role of aluminum in glasses.

7 Covalent, Non-Oxidic Glasses

Non-oxide glasses formed by the heavier chalcogenides and pnictides of Main Group III-V elements are a promising new class of solid state materials with intriguing optical and electrical properties. On the basis of extensive spectroscopic and theoretical structural studies during the past decade, general structural concepts for these glasses are now emerging slowly. New conceptual approaches have addressed the question of short-range order on the basis of average valences [143] and coordination numbers [144], while the theoretical framework of rigidity percolation [145] has been applied to explain the macroscopic properties of these materials [146]. Experimentally, EXAFS and diffraction studies appear to indicate substantial intermediate-range order on the 10–20 Å scale in many covalent glasses [3].

In general, the types of non-oxide glass systems currently under investigation can be grouped into two classes: (a) infrared transparent materials based on binary and ternary III–V chalcogenides or post-transition metal phosphides and arsenides, and (b) fast ion conducting glasses, based on stoichiometric Main Group III–V chalcogenides and stoichiometric sulfides of electropositive cations.

7.1 *Glassy and Crystalline Silicon Chalcogenides*

In contrast to SiO_2, the heavier silicon chalcogenides SiS_2 and $SiSe_2$ can only be converted to the glassy state by rapid melt-quenching. Extensive infrared and Raman spectroscopic studies carried out on these glasses have revealed the presence of $SiX_{4/2}$ tetrahedra interconncted by edge-sharing [147]. Such edge-sharing units are well-recognized structural features in the crystal chemistry of silicon sulfides and selenides [148]. However, in the glassy state they constitute a violation of the traditional Zachariasen network model, which explicitly excludes any connectivities different from corner-sharing. While vibrational spectroscopy is limited to the qualitative aspect, further quantitation can be obtained by solid state NMR [149–151]. In contrast to the ^{29}Si MAS-NMR spectrum for glassy SiO_2, the spectra of glassy SiS_2 and $SiSe_2$ show multiple peaks, suggesting that the bond angle distribution functions in glassy SiS_2 and $SiSe_2$ are non-monotonic and contain distinct maxima and minima. The three broad maxima observed in the NMR spectra can be assigned to silicon tetrahedra sharing edges with no, one or two adjacent tetrahedral units $E^{(0)}$, $E^{(1)}$, and $E^{(2)}$, respectively, as shown in Fig. 35. Evidently, the individual $E^{(n)}$ species have sufficiently different average Si–S–Si bond angles, so that discrete resonances can be observed.

Figure 36 shows very recent results in the Si–S system. Consistent over multiple sample preparations, the spectrum of bulk SiS_2 glass shows that the resonance attributed to $E^{(2)}$ units is actually split into two components near −17 and −21 ppm, respectively. Based on the chemical shift of crystalline SiS_2, whose structure is composed entirely of $E^{(2)}$ units, the −21 ppm resonance in the glass is assigned to Si atoms within longer chains, where $E^{(2)}$ units are adjacent to only other $E^{(2)}$ units. The −17 ppm resonance is attributed to $E^{(2)}$ units that are part of smaller chain fragments, where $E^{(2)}$ units are adjacent to $E^{(1)}$ units. Figure 36 also includes NMR results on some newly prepared amorphous SiS_x:H films ($x \sim 2$), synthesized by plasma-enhanced chemical vapor deposition (PECVD) from SiH_4 and H_2S. Compared to the bulk glass, the PECVD samples are seen to possess higher concentrations of $E^{(0)}$, and reduced concentrations of $E^{(2)}$ species, particularly of those contained within longer chains. This structural speciation is critically influenced by the H_2S/SiH_4 flow rate ratio under synthesis conditions. Thus, PECVD offers the opportunity of controlling the ratio of corner-shared to edge-shared $SiS_{4/2}$ tetrahedra in the final product [151]. This situation is different from the molten state synthesis,

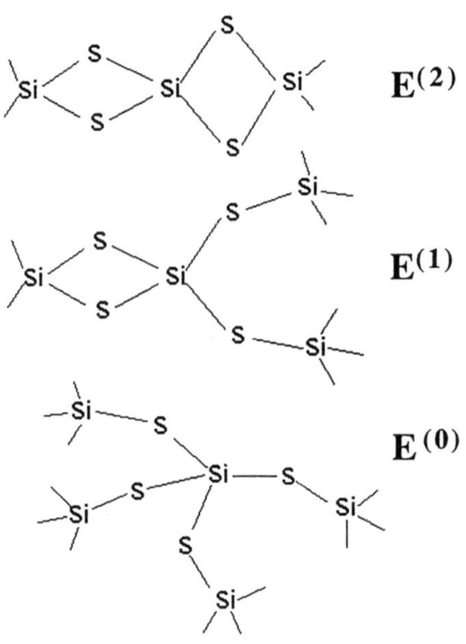

Fig. 35. Possible sites in silicon chalcogenide glasses and their nomenclature

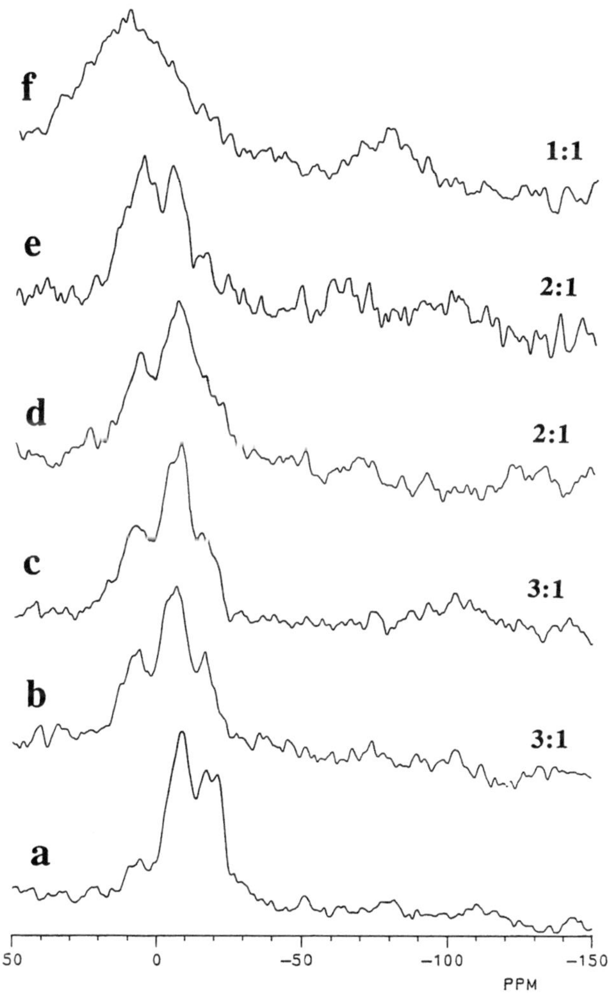

Fig. 36a–f. ^{29}Si MAS-NMR spectra of samples in the Si—S system. **a** bulk glassy SiS_2 prepared from the molten state. **b–f** amorphous SiS_x:H samples prepared by PECVD as a function of H_2S/SiH_4 flow rate ratio (indicated on the figure). Reproduced from Ref. [151]

where changes in the silicon/sulfur melt composition just produce different relative amounts of segregated phases consisting of glassy SiS_2 and elemental sulfur [152].

Figure 37 shows most recent results in the silicon-selenium system, where two new low-temperature crystalline phases have been characterized recently [153]. The ^{29}Si MAS-NMR spectra of these phases suggest that the structure of glassy $SiSe_2$ closely resembles that of a low-temperature crystalline $SiSe_2$ phase stable near 500 °C, close to the temperature, where molten $SiSe_2$ undergoes its glass transition. Si-Se glasses with higher Se contents show an increase of the $E^{(0)}$ component, whereas the $E^{(2)}$ component is essentially absent [150]. The same result is obtained in mixed $SiSe_2$-P_2Se_5 glasses, indicating a more

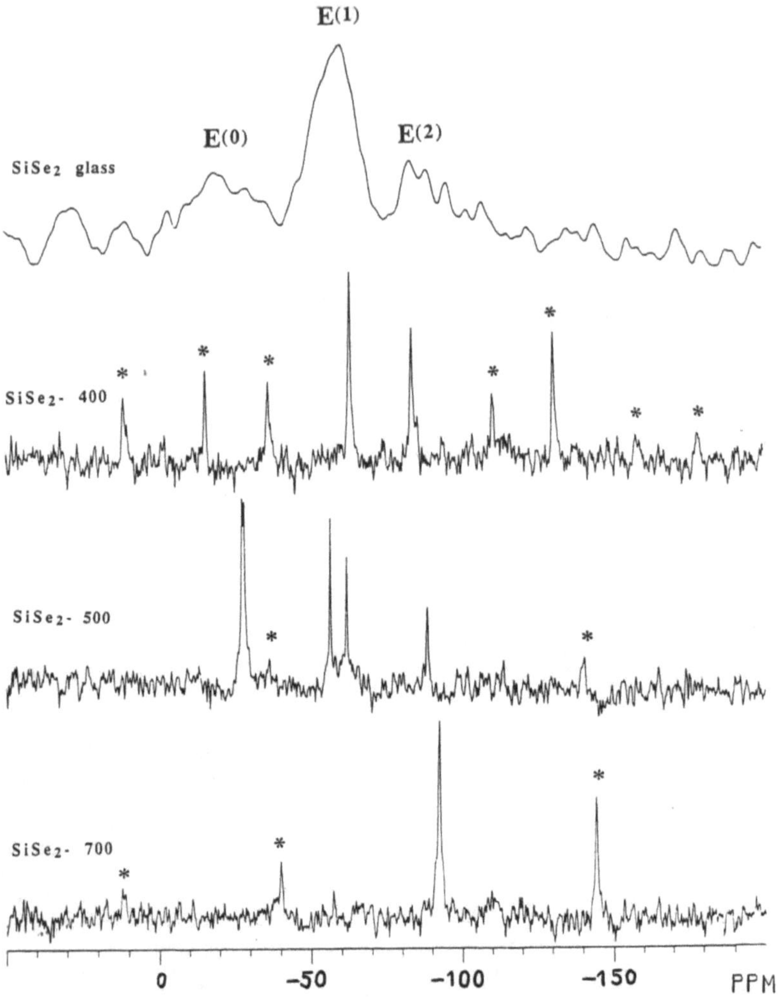

Fig. 37. ^{29}Si MAS-NMR spectra of samples in the Si—Se system: *top*: glassy $SiSe_2$. The other three spectra correspond to three different crystalline modifications of $SiSe_2$. Spinning sidebands are indicated by *asterisks*. Reproduced from Ref. [153]

continuous distribution of silicon microstructures, including the presence of Si—Se—Se and Si—Se—P linkages [150].

7.2 The System Phosphorus-Selenium

The system Phosphorus-Selenium is characterized by a pronounced tendency towards vitrification. Glasses are formed over a wide compositional region, from 0–52 and 62–80 mol% P. In part due to mutually contradictory phase diagrams, the structural organization of these glasses has been subject to a great deal of speculation and controversy [154–158]. Extensive spectroscopic investigations have been carried out [159–161], but it has been only very recently that modern solid state NMR techniques have been employed [162–167]. Figure 38 shows some possible short-range order environments. While MAS-NMR chemical shifts clearly distinguish between four-coordinate P atoms (Se = $PSe_{3/2}$) and three-coordinated P atoms they do not allow one identify the number of P or Se nearest neighbors unambiguously [163].

The latter question has been successfully addressed by a detailed analysis of the spin echo intensity as a function of evolution time. This experiment provides quantitatively reliable information on the strength of the ^{31}P–^{31}P dipole–dipole coupling, as verified with a range of crystalline model compounds. Due to the substantial difference in the strength of the homonuclear dipole-dipole interactions, the spin echo decays for P atoms coordinated to selenium only and those for the P atoms with a phosphorus nearest neighbor are distinctly different. Typical simulations, modelling the contributions for both types of species in the glasses are shown in Figure 39. Indeed, the experimental ^{31}P spin echo decay functions in the P—Se glasses reveal that P—P bonds contribute above 25 at.% P, and their contribution can be quantitatively evaluated. The combined analysis of ^{31}P spin-echo and MAS-NMR results then yields the final phosphorus speciation shown in Figure 40. [166]. There is a definite preference towards formation of P—Se rather than P—P bonds. Likewise Se—P bonds are greatly favored over Se—Se bonds, although the latter can be found over the entire glass-forming region. While there exists a distinct driving force for the formation of $PSe_{3/2}$ groups, the reaction of these units with

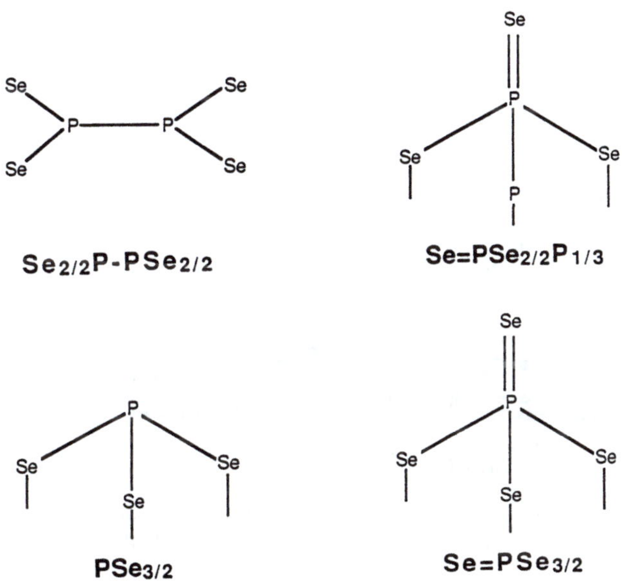

Fig. 38. Possible local environments in Phosphorus-Selenium glasses

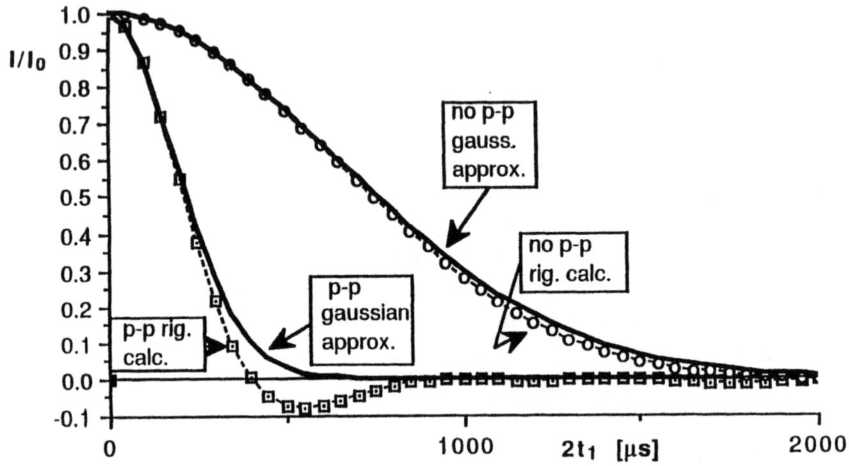

Fig. 39. Computer simulations of spin-echo decays for P atoms bonded to one P and for P atoms not bonded to P in a P—Se glass containing 40 at.% P randomly distributed beyond the first coordination sphere. Included are the results of a rigorous calculation and a Gaussian approximation

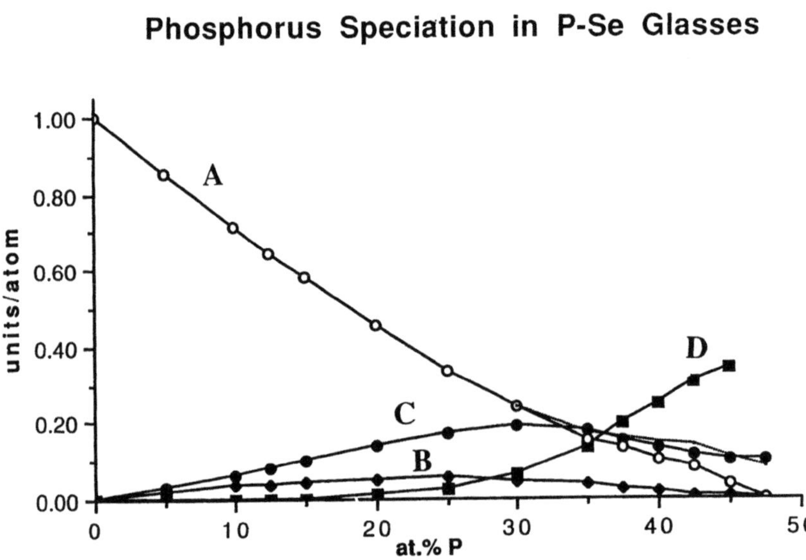

Fig. 40. Site speciation in P—Se glasses, as deduced from the combined analysis of spin echo and MAS-NMR investigations. *A*: Se—Se bonds, *B*: $PSe_{3/2}$ units, *C*: $Se{=}PSe_{3/2}$ units, *D*: P—P bonded units. Reproduced from Ref. [166]

excess Se to form $Se = PSe_{3/2}$ groups is much less favored. The multitude of structural units present at each composition can be visualized to disfavor nucleation processes that would eventually lead to crystallization.

It is important to realize that all of the above structural findings, obtained on melt-quenched glasses, reflect molten-state chemical equilibria that are frozen in at the glass transition temperature, T_g. To understand these speciations further, and to predict the influence of quenching rates on glass structures, it is important to characterize these equilibria and the thermodynamic parameters involved. Consider, for instance the compositional dependence of the P^{IV} mole fraction, i.e. the fraction of $Se = PSe_{3/2}$ groups in the P—Se system, replotted in Fig. 41. The data can be consistently described assuming a melt-equilibrium model:

$$PSe_{3/2} + Se \Leftrightarrow Se = PSe_{3/2}$$

adopting an equilibrium constant around 0.8 $at.^{-1}$ Recent in-situ high-temperature NMR studies have provided additional insights into this equilibrium

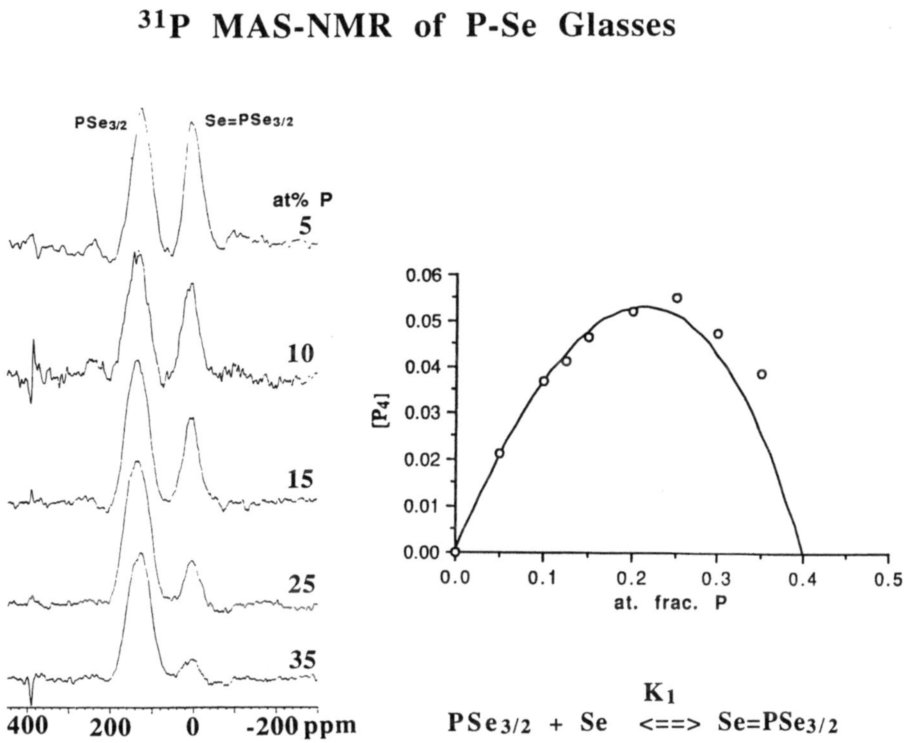

Fig. 41. Left: ^{31}P MAS-NMR spectra of P—Se glasses with various P contents. Right: Concentration of four-coordinate $Se{=}PSe_{3/2}$ units in binary P—Se glasses. The *solid curve* is calculated assuming an equilibrium of the kind $PSe_{3/2} + Se \Leftrightarrow Se{=}PSe_{3/2}$ with an equilibrium constant K_1 of 0.83 $at.^{-1}$

[168,169]. Approximately 200 K above T_g the NMR spectra are in the fast chemical-exchange limit, and the chemical shift then reflects the average of the phosphorus site populations.

$$\delta_{ave} = N_4\delta_4 + (1 - N_4)\delta_3 \tag{14}$$

where N_4 is the fraction of four-coordinated P atoms and δ_3 and δ_4 are the chemical shifts of the individual resonances contributing to the exchange equilibrium.

A detailed inspection of the chemical shift as a function of temperature and composition in terms of Eq. (14) reveals that N_4 consistently decreases with increasing temperature, reflecting a decrease in K_1. A corresponding van't Hoff plot is shown in Fig. 42, from which the enthalpy and entropy of reaction (14) can be estimated. Especially noteworthy is the good agreement of this analysis with the K_1 values at the respective T_g values, determined independently by room-temperature MAS NMR on the glassy samples.

A second temperature-dependent chemical equilibrium is apparent in glasses containing ≧40 at.% P, resulting in the appearance of molecular P_4Se_3 cluster units, whose spectral contribution grows with increasing temperature. In essence, the process monitored here is a depolymerization reaction, which has been modeled in terms of the scheme

$$3\ Se_{2/2}P{-}PSe_{2/2} \rightarrow P_4Se_3 + 2\ PSe_{3/2} \tag{15}$$

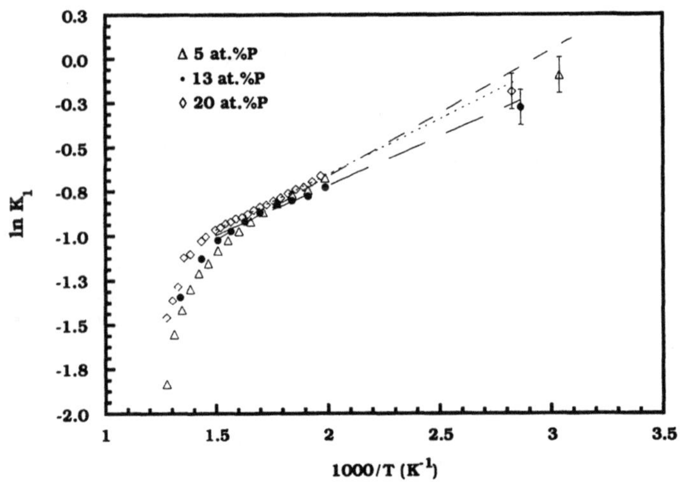

Fig. 42. Temperature dependence of K_1 describing the equilibrium $PSe_{3/2} + Se \Leftrightarrow Se{=}PSe_{3/2}$, as extracted from temperature dependent chemical shift measurements in the fast exchange limit. Reproduced from Ref. [169]

The temperature dependence of the associated equilibrium constant indicates a reaction enthalpy of 50 kJ/mol. Beyond ca. 400 °C the spectra evidence chemical exchange between the P_4Se_3 molecules and the residual P atoms in the molten matrix. Figure 43 illustrates that the experimental data can be simulated well by a three-site exchange model. An Arrhenius plot of the exchange rate constants thus extracted yields an activation energy of 83 kJ/mol [168].

Complementary thermodynamic and kinetic data have been obtained using variable temperature ^{77}Se NMR studies, the results of which are shown in Fig. 44. [169]. At low temperatures, distinct resonances for P—bonded and Se-only bonded Se atoms can be resolved, which coalesce approximately 100 K above T_g into a single resonance due to chemical exchange on the NMR timescale. At higher temperatures, the average ^{77}Se chemical shift reflects the quantitative distribution of Se—Se versus Se—P bonding in the glass; tentative speciations obtained from such data differ somewhat from the numbers obtained from ^{31}P—^{77}Se SEDOR NMR [45], but show substantially the same trend with composition. Overall, neither the high-temperature nor the SEDOR data provide strong evidence for phosphorus clustering.

7.3 Ternary Phosphorus Chalcogenide Systems

Introduction of a third glass constituent such as arsenic, tellurium, germanium, or thallium into P-Se based glasses has a profound influence on the thermal properties and the chemical stability. Extensive ^{31}P MAS-NMR studies have been carried out to elucidate the structural effects of such compositional modifications. Partial substitution of P by As within the compositional series $[P_xAs_{1-x}]_{1-y}Se_y$ leads to a systematic increase in N_4 [170]. This increase has been modeled by assuming an intermolecular stabilization of $Se = PSe_{3/2}$ groups with $AsSe_{3/2}$ units, an association equilibrium of the form

$$Se = PSe_{3/2} + AsSe_{3/2} \Leftrightarrow Se_{3/2}P = Se \cdots\cdots AsSe_{3/2} \tag{16}$$

characterized by an equilibrium constant K_2.

As illustrated by Fig. 45, the experimental phosphorus speciations can be modeled in their entirety with a single set of equilibrium constants $K_1 = 0.8$ and $K_2 = 17$, thus lending strong credibility to this model [171]. Alloying binary P—Se glasses with tellurium and thallium also serves to stabilize the four-coordinated P atoms, while alloying with germanium was observed to produce the opposite effect [172].

7.4 The Phosphorus-Sulfur System

Glasses in the system P-S have been investigated over the range $0 \leqq$ at.% $P \leqq 25\%$, using solution state ^{31}P NMR on CS_2-extracts [173, 174], and ^{31}P

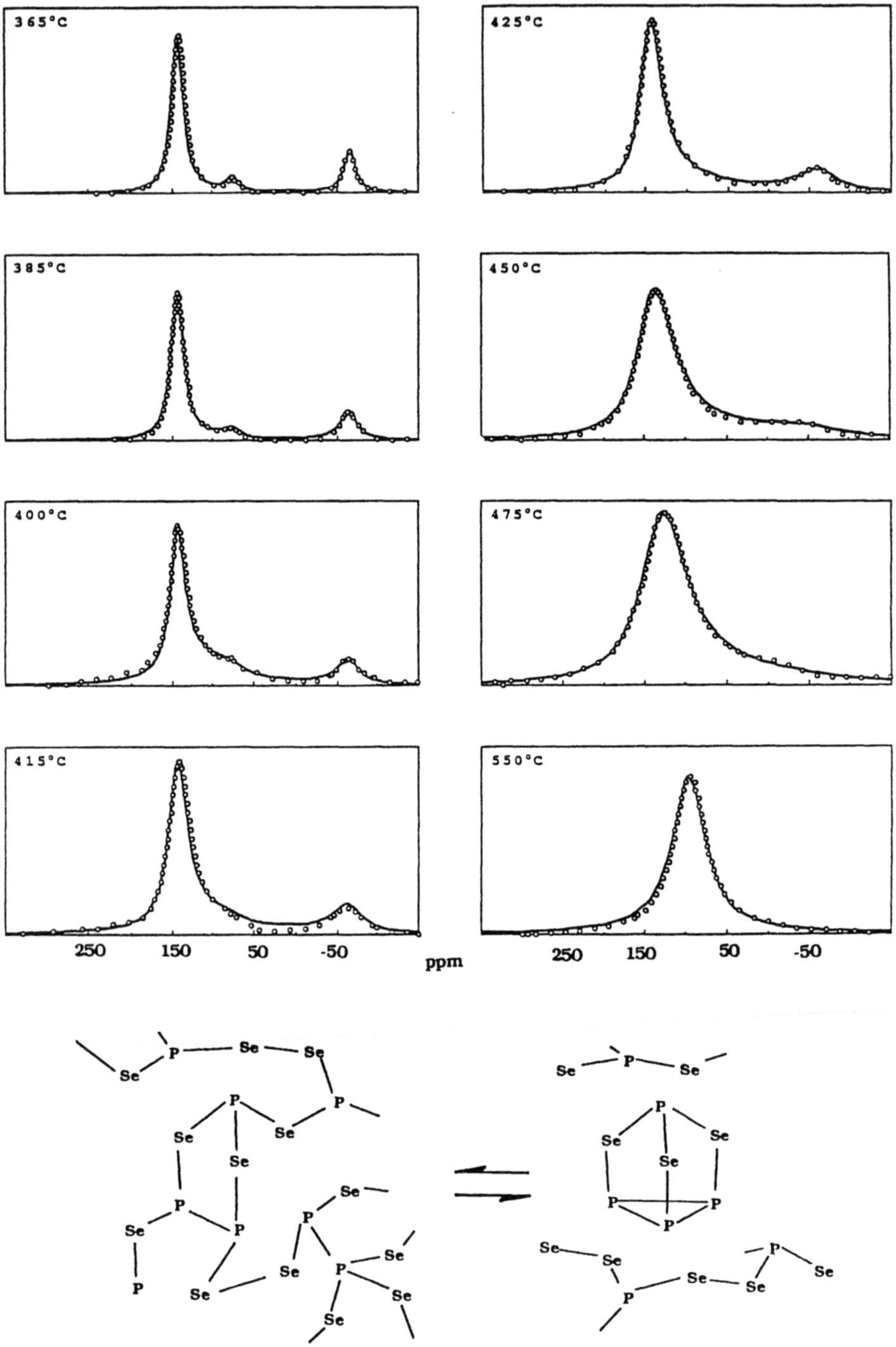

Fig. 43. Experimental (*circles*) and simulated ^{31}P lineshapes as a function of temperature, revealing the manifestations of chemical exchange between molecular P_4Se_3 (peaks at 70 and -70 ppm) and the remaining P in the melt (peak at 140 ppm). The chemical exchange process is sketched below. Reproduced from Ref. [168]

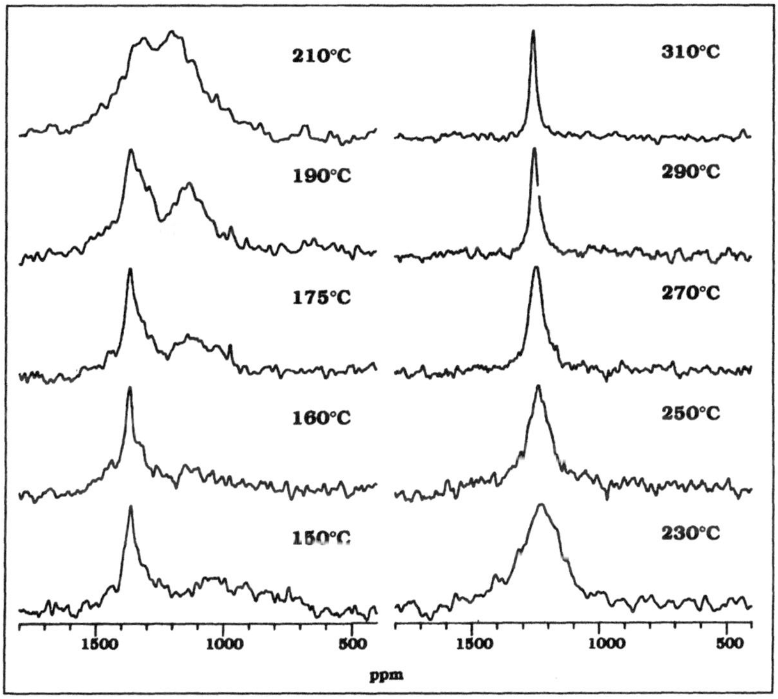

(a)

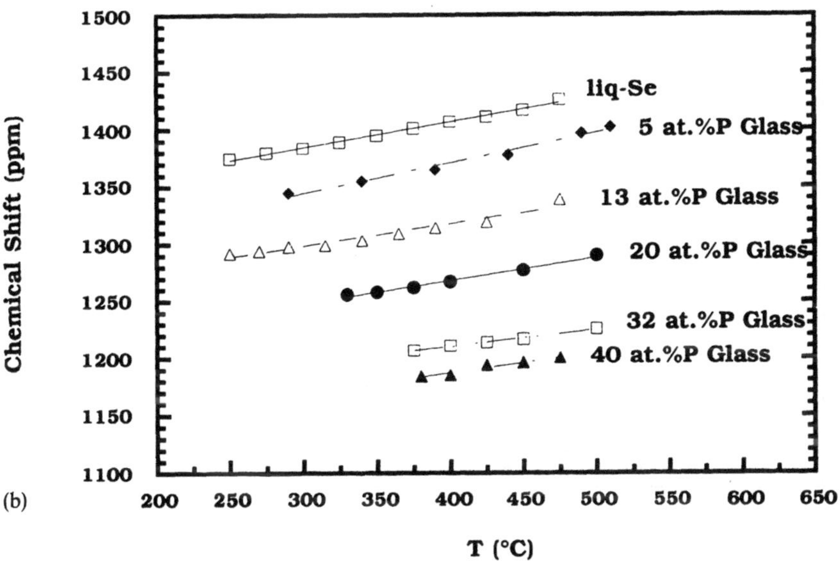

(b)

Fig. 44a. Temperature dependent ^{77}Se NMR spectra of a P–Se glass containing 20 at. % P. The resonances observed at low temperatures near 1380 and 1150 ppm correspond to Se-only bonded and P-bonded Se atoms, respectively. Reproduced from Ref. [169]. **b** Temperature dependence of ^{77}Se chemical shifts in various P-Se melts. Reproduced from Ref. [169]

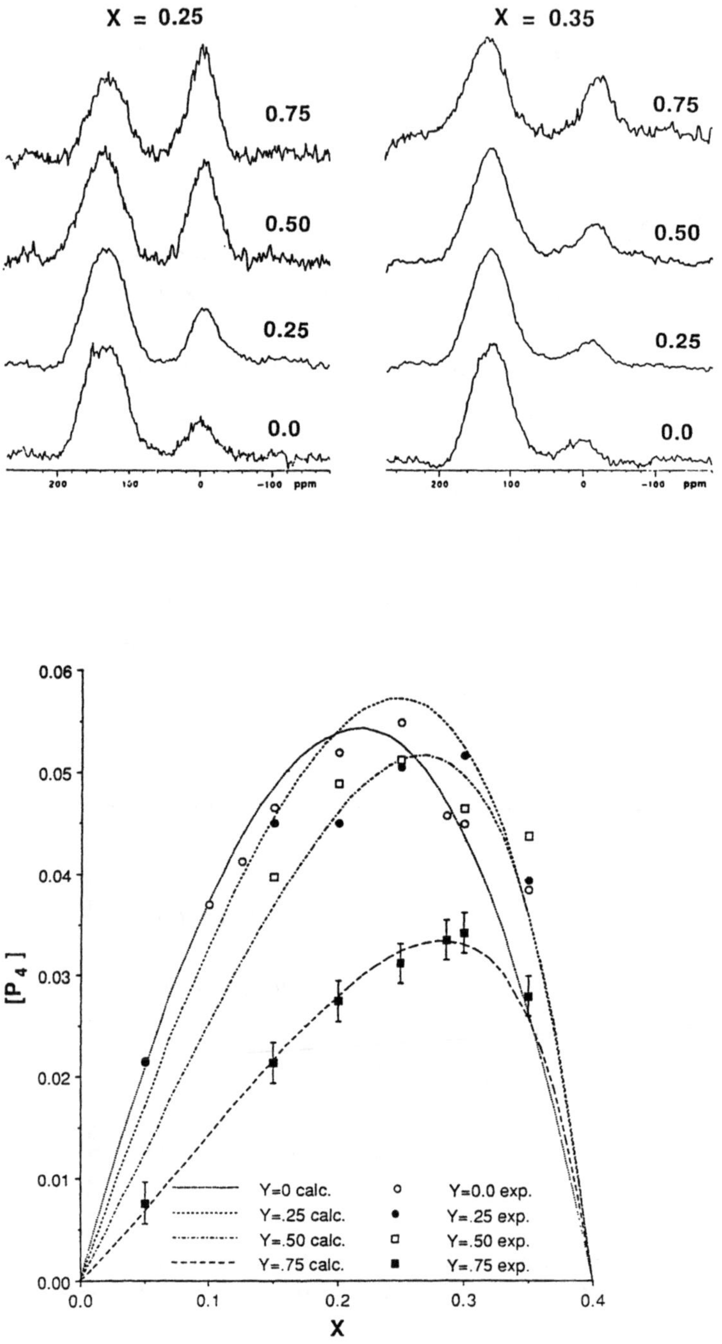

Fig. 45. ^{31}P MAS-NMR results on glasses with composition $(P_{1-y}As_y)_xSe_{1-x}$. *Top*: typical experimental spectra; *bottom*: concentration of $Se{=}PSe_{3/2}$ groups as a function of x and y. *Dashed dotted*, and *solid curves* correspond to calculations described in the text. Reproduced from Ref. [171]

wideline, MAS, and spin echo NMR techniques [175], in conjunction with parallel studies of crystalline model compounds [176]. For glasses with <15 at.% P the ^{31}P chemical shift tensor reveals near-axial symmetry, and its isotropic value (113 ppm) and principal tensor components support the presence of $S=PS_{3/2}$ rather than of $PS_{3/2}$ groups. At P contents above 15 at.% P the formation of P_4S_9 and P_4S_{10} clusters is apparent in the spectra, giving rise to sharper resonances around 51 and 57 ppm. Spin echo NMR experiments suggest that the dipole-dipole coupling is slightly stronger than for a random distribution of $S=PS_{3/2}$ groups, hence also supporting the formation of clusters. No P-P bonded species are detected. The MAS-NMR spectrum of glassy $AsPS_4$

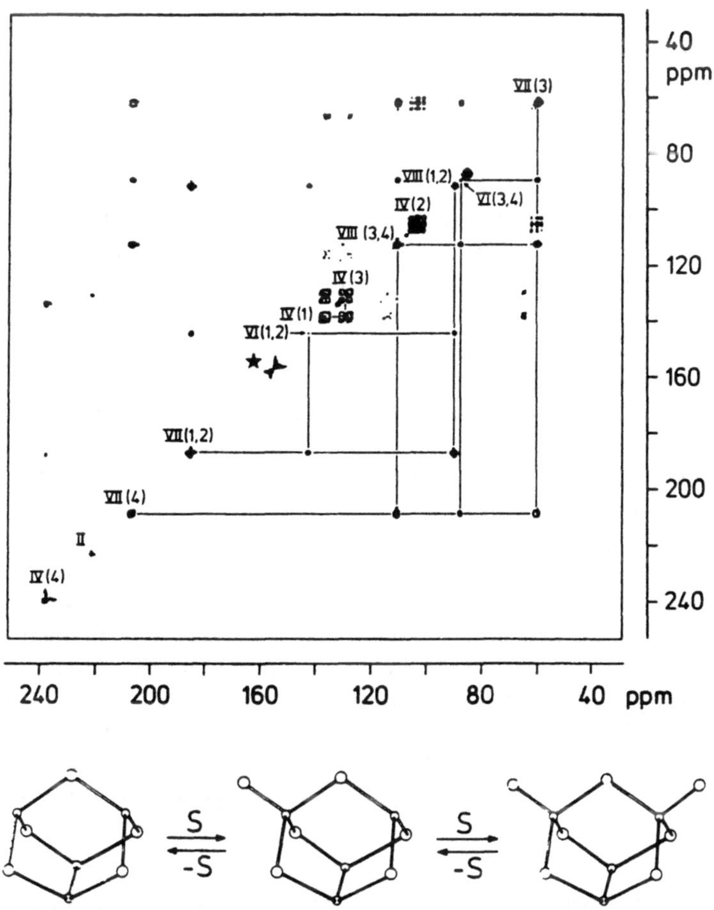

Fig. 46. 2-D NOESY-NMR data on a P—S melt containing 44.4 mol% phosphorus. The crosspeaks suggest the existence of chemical exchange between P_4S_5, P_4S_6 and P_4S_7 molecules, which exist as discrete units within the melt. Diagonal peaks are the individual sites of various P_4S_x molecules. Reproduced with permission from Ref. [24]

resembles that of binary P-S glasses at low P contents and has therefore been assigned to $S = PS_{3/2}$ groups [177]. High-resolution liquid state NMR techniques reveal that molten phosphorus-sulfur melts consist essentially of discrete P_4S_x molecules, and very little polymerization takes place. Figure 46 shows a two-dimensional exchange NMR spectrum of a sample with nominal stoichiometry P_4S_5, revealing the presence of chemical exchange between molecular P_4S_5, P_4S_6, and P_4S_7, respectively, following the scheme illustrated.

7.5 Covalent Pnictide Glasses

A variety of systems containing phosphides and arsenides of Main Group and post transition elements, most notably $CdGeAs_2$, $CdGeP_2$ and analogous alloys with the II-IV-V_2 stoichiometry can be prepared in the glassy state by rapid melt-quenching [179]. Most strikingly, these glasses have consistently higher densities than the corresponding crystalline compounds, hence suggesting that there are marked structural differences between these two states of matter [180]. Crystalline $CdGeAs_2$ and $CdGeP_2$ are strictly chemically ordered with As^{3-} and P^{3-} ions occupying exclusively the anionic sites of the tetragonal chalcopyrite structure, which is essentially a distorted heteropolar analog of the diamond (or silicon) lattice. The issue of glass-formation in such systems, in which all of the atoms are tetrahedrally coordinated, has been subject of considerable research activity. Molecular modeling has been used to "amorphize" the cubic silicon lattice by the introduction of random bond switches, which alter the ring statistics of the original lattice and newly introduce five- and seven-membered rings [181, 182]. For heteropolar structures such as III–V or II–IV–V_2 lattices the presence of such odd-membered rings necessarily implies the formation of homopolar (such as V—V) bonds. Past attempts to identify such "wrong bonds" in amorphous GaAs by ^{69,71}Ga and ^{75}As NMR have only met with limited success, because the quadrupolarly perturbed line shapes for these nuclei are difficult to interpret [183]. In contrast, the $CdGeP_2$–$CdGeAs_2$ system is a III–V analog containing two spin-1/2 isotopes well suited for NMR studies. In particular, the internuclear dipole-dipole couplings, due to their distance dependence are highly sensitive to the formation of homopolar bonds. With this objective, the chemical bond distribution in glassy $CdGeP_2$–$CdGeAs_2$ alloys has been examined in detail using ^{31}P spin echo and ^{31}P-^{113}Cd SEDOR NMR approaches [184, 185]. Figure 47 shows the comparison of the experimental ^{31}P spin echo decay data with simulations for a chemically ordered

Fig. 47. ^{31}P spin echo NMR results in $CdGeAs_{2-x}P_x$ glasses. The experimental data are compared with simulations corresponding to a random distribution of P and As atoms over the anionic sites of the chalcopyrite lattice ("ordered model", *solid curves*), and to a random distribution of P and As over all sites of a zincblende lattice ("random model", *dashed curves*). The *dotted curves* show the linear combination of both models that reproduces best the experimental data. Reproduced from Ref. [185]

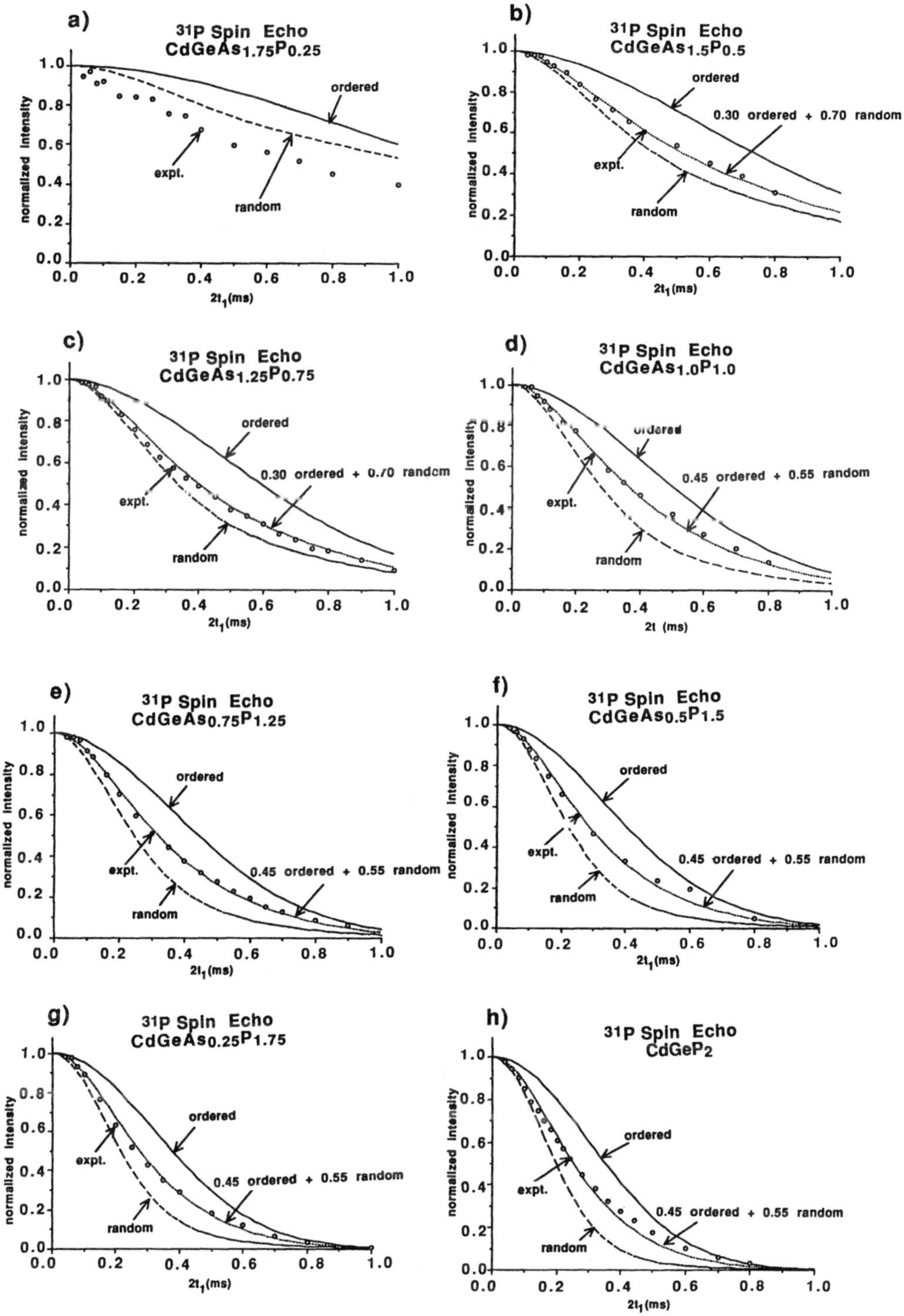
a)
31P Spin Echo
CdGeAs1.75P0.25
ordered
expt.
random
normalized intensity
2t1(ms)
b)
31P Spin Echo
CdGeAs1.5P0.5
ordered
0.30 ordered + 0.70 random
expt.
random
normalized intensity
2t1(ms)
c)
31P Spin Echo
CdGeAs1.25P0.75
ordered
0.30 ordered + 0.70 random
expt.
random
normalized intensity
2t1(ms)
d)
31P Spin Echo
CdGeAs1.0P1.0
ordered
0.45 ordered + 0.55 random
expt.
random
normalized intensity
2t (ms)
e)
31P Spin Echo
CdGeAs0.75P1.25
ordered
0.45 ordered + 0.55 random
expt.
random
normalized intensity
2t1(ms)
f)
31P Spin Echo
CdGeAs0.5P1.5
ordered
0.45 ordered + 0.55 random
expt.
random
normalized intensity
2t1(ms)
g)
31P Spin Echo
CdGeAs0.25P1.75
ordered
0.45 ordered + 0.55 random
expt.
random
normalized intensity
2t1(ms)
h)
31P Spin Echo
CdGeP2
ordered
0.45 ordered + 0.55 random
expt.
random
normalized intensity
2t1(ms)

chalcopyrite structure, a random distribution of P atoms over all the lattice sites, and an additive superposition of both. The experimental results are quantitatively consistent with the occurrence of P—P bonds in glassy $CdGeP_2$, albeit at a sub-statistical level. In fact, Fig. 47 suggests that partial site randomization appears to occur within the entire $CdGeP_2$–$CdGeAs_2$ compositional region.

Figure 48 shows results from ^{31}P-^{113}Cd SEDOR NMR for selected compositions in this system. The experimental SEDOR decays are compared with the predictions from a chemically ordered model, in which each Cd atom is surrounded by four pnictogen atoms (the P fraction of which depending on the batched P/As ratio) and a completely random distribution of Cd and P relative to each other. Clearly, the SEDOR data support the random model, suggesting that the average number of Cd—P bonds is lower than in the crystalline state. As a caveat, it must be noted, however, that the analysis of these SEDOR data

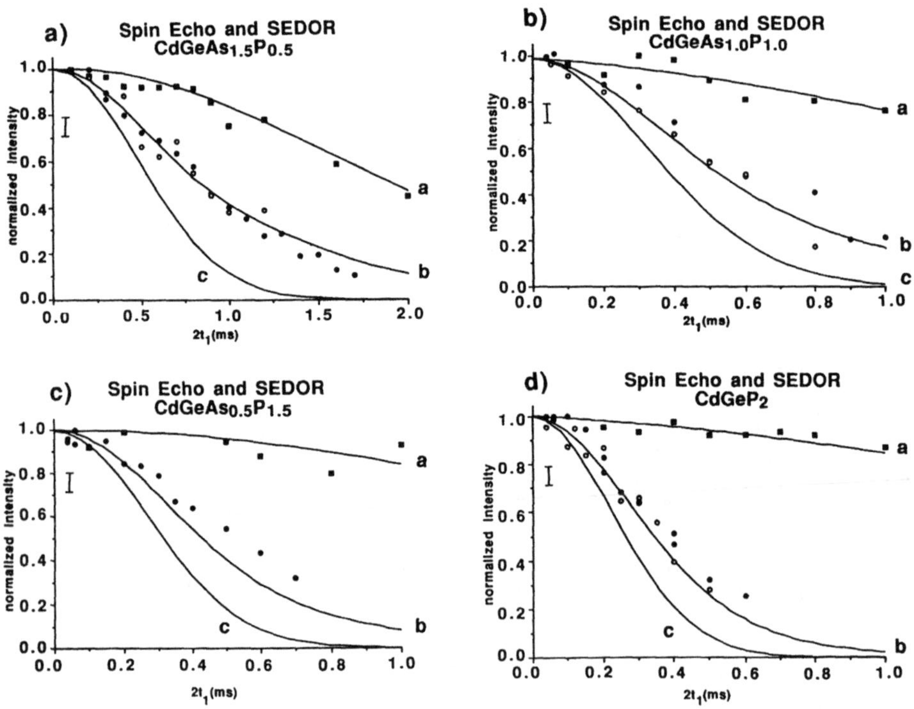

Fig. 48a–d. Experimental and calculated ^{31}P—^{113}Cd spin echo double resonance NMR results in $CdGeAs_{2-x}P_x$ glasses. *Open* and *filled circles* correspond to two independently obtained datasets. The experimentally measured ^{113}Cde spin echo decays are also include (*filled squares*). The *solid curves* show: **a** polynomial fit to the ^{113}Cd spin echo decay data, **b** simulations based on a random distribution of P relative to Cd (P and Cd distributed randomoly over all sites in the chalcopyrite lattice), **c** simulations based on an ordered distribution of P relative to Cd (P and As randomly distributed over the anionic sites only in the chalcopyrite lattice). Reproduced from Ref. [185]

does not take into account possible contributions of indirect $^{113}Cd—^{31}P$ dipole–dipole interactions (and their anisotropies).

Disparity between glassy and crystalline structures is also evident in the ^{113}Cd spin echo NMR studies of compositions in the system $CdGe_xAs_2$ ($0 \leqq x \leqq 1$) [186]. Here, the spin echo decays of glassy $CdAs_2$ and $CdGeAs_2$ are substantially more rapid than those of their crystalline counterparts, and rather resemble that observed for crystalline Cd_3As_2. The suggestion that the structure of these glasses is characterized by the presence of Cd_3As_2 and As domains is consistent with the metastable eutectic observed in the Cd—As system at 61.5% As [187]. In excellent agreement with this prediction, most recent DSC and X-ray studies have revealed that glassy $CdAs_2$ undergoes either a glass → glass transition or a primary crystallization step to form Cd_3As_2 and As, before converting into the equilibrium phase $CdAs_2$ [188, 189].

7.6 Ionically Conductive Li- and Ag- Chalcogenide Glasses

These types of chalcogenide glasses are compositionally analogous to the usual oxide glass systems discussed above, differing from the latter merely in the replacement of oxygen by sulfur or selenium. Experimental studies have concentrated on Li_2S containing glasses based on B_2S_3, SiS_2, and P_2S_5, which are formed by rapid quenching techniques [190]. Although these materials are solid electrolytes with high room temperature ionic conductivities (10^{-2} S/cm), widespread applications of these glasses as components in lithium batteries have been limited, perhaps due to their instability in moist air, their recrystallization above T_g, and their reactivity towards lithium metal anodes. The development of new materials with improved thermal and chemical stabilities without compromising the ionic conductivity is an area of active research, and requires support by structural investigations. In this context, a salient question is in how far the compositional analogy between these glasses and binary alkali borate, silicate, and phosphate glasses translates into structural analogy.

7.6.1 Metal Thioborate Glasses

Like its oxide analog, glassy B_2S_3 is known to accommodate network modifiers such as Li_2S, Na_2S, and Tl_2S, although fast quenching (ice-water) of small samples is generally necessary to preserve the glassy state. ^{11}B wideline and MAS NMR have confirmed that four-coordinated boron atoms are generated in these glasses, analogous to the situation in binary borate glasses. Figure 49 summarizes the results obtained so far on various systems [191–196]. It is noteworthy that specifically in the regions of low sulfide modifier concentrations the fraction of four-coordinate boron atoms, N_4, is substantially higher than in related oxide glass systems. These findings are possibly related to the formation of structural environments in the glasses that resemble those found

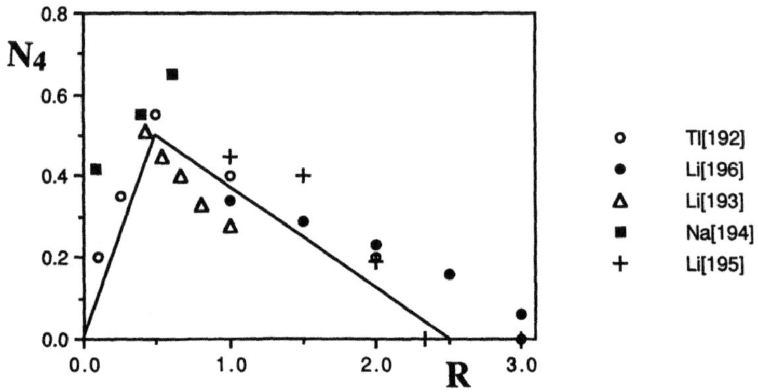

Fig. 49. ^{11}B–N_4 data plotted versus R = mol% M_2S/mol% B_2S_3 in various metal thioborate glass systems. The *solid curve* indicates the known behavior for many alkali metal borate systems

in various crystalline silver and lithium thioborates in which all of the boron atoms are four-coordinated [197]. The oxide phases in the corresponding compositional region have completely different structures. In conjunction with the conclusion that boron chalcogenide melts are essentially disordered assemblies of microstructures known to exist in the crystalline phase diagram, the structural differences between boron oxide and boron sulfide based glasses are thus easily understood.

7.6.2 Metal Thio- and Selenosilicate Glasses

Glasses in the system $(Li_2S)_x$-$(SiS_2)_{1-x}$ ($0.4 \leqq x \leqq 0.6$) have been studied as interesting test cases on whether the compositional analogy of these glasses to binary alkali silicate systems translates into structural analogy as well. This question has been difficult to examine, however, because the ^{29}Si chemical shifts are dominated by the $E^{(n)}$ rather than the $Si^{(n)}$ specification; this means also that the number of non-bridging sulfur atoms produces only minor chemical shift changes for a given $E^{(n)}$ species [178, 179]. It is therefore not possible to quantitate the number of non-bridging sulfur atoms. The NMR spectra show, however, that introduction of Li_2S into glassy SiS_2 initially reduces the extent of edge-sharing markedly, and the $E^{(2)}$ species have disappeared for the composition $x = 0.30$. Figure 50 illustrates the network modification scheme developed on the basis of the NMR data. This model reveals that, implicitly, the destruction of edge-sharing units by Li_2S (represented by diagonal arrows) will result in the formation of non-bridging sulfur atoms. In the compositional region where the most stable glasses form ($0.38 < x < 0.6$) the $E^{(n)}$ distribution remains approximately constant, with a $E^{(0)}/E^{(1)}$ ratio of ca. 2.5:1. The structure of Na_2S-SiS_2 glasses is essentially governed by the same principles as discussed above, although the fraction of edge-sharing units appears to be consistently

(a)

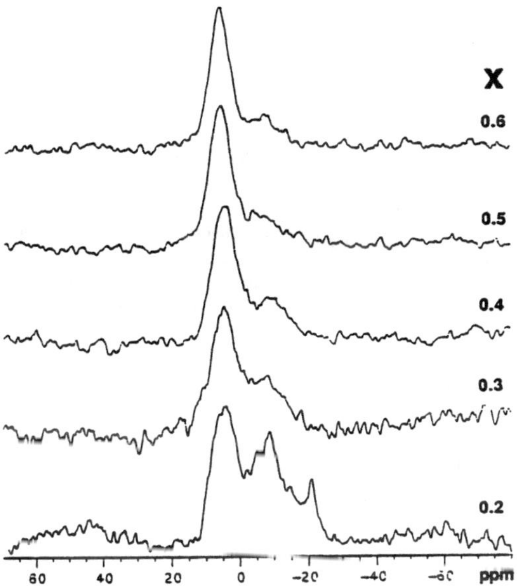

(b)

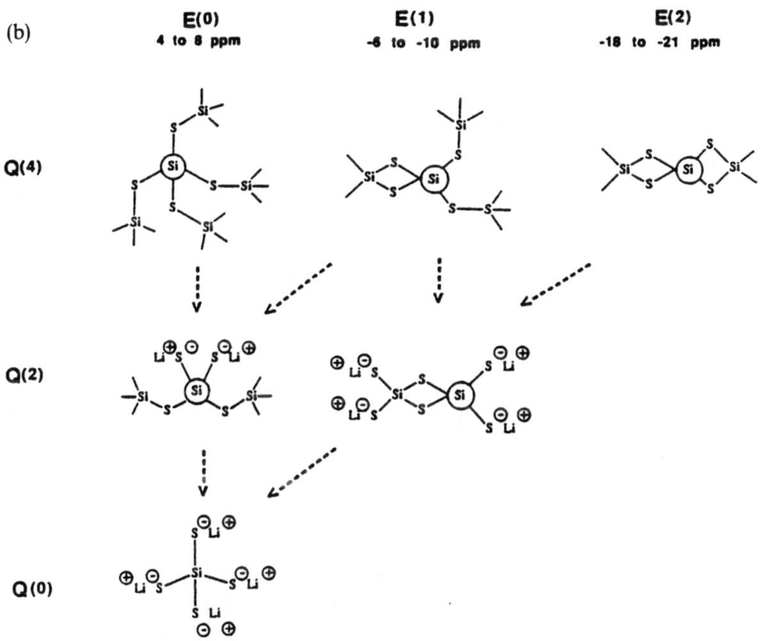

Fig. 50a, b. ^{29}Si MAS-NMR spectra **a** and structural interpretation **b** for glasses in the $(Li_2S)_x$ $(SiS_2)_{1-x}$ system

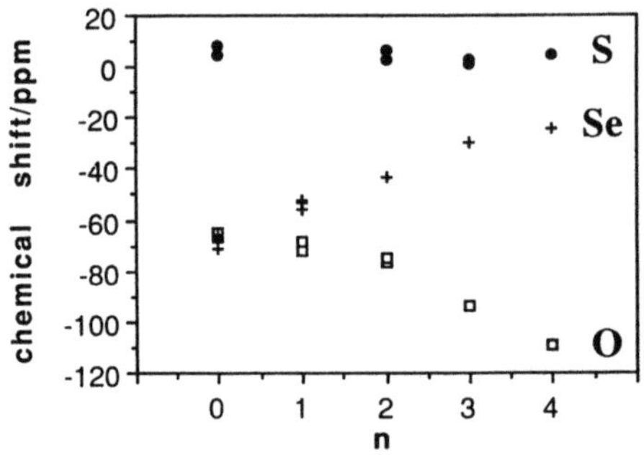

Fig. 51. The dependence of the ^{29}Si chemical shift on n in the various $Si^{(n)}$ units for lithium silicon oxide, sulfide, and selenide systems. The data shown for the selenide system must be considered as tentative, because the structural assignment is based on sample stoichiometry only, due to the lack of crystallographic data

higher [200]. The additional presence of Al_2S_3 in these glasses tends to diminish the $E^{(1)}$ units further [201]. Finally, extension of these investigations to the system Li_2Se-$SiSe_2$ reveals qualitatively many of the same features, except that the chemical shift discrimination between various $Si^{(n)}$ units is markedly improved here, and semi-quantitative species concentrations can be derived from the ^{29}Si MAS-NMR spectra [202]. Parallel studies of crystallized samples reveal that with increasing number of bridging selenium atoms in the $Si^{(n)}$ notation the chemical shifts become less negative, the opposite trend as observed in the silicates [202] (see Fig. 51).

7.6.3 Metal Thiophosphate Glasses

^{31}P MAS-NMR studies have been undertaken on a variety of thiophosphate glasses, including the binary systems Li_2S–P_2S_5 [203] and Ag_2S–P_2S_5 [204], and the ternary systems Li_2S–SiS_2–P_2S_5 [205], Li_2S–B_2S_3–P_2S_5 [205] and Li_2S – Al_2S_3–P_2S_5 [206]. In many of these glasses, large chemical shift differences between crystalline and glassy phases are observed.

For instance, glasses with the composition $Li_4P_2S_7$ disproportionate on crystallization, according to

$$Li_4P_2S_7\ (\text{glass}) \rightarrow Li_4P_2S_6\ (\text{cryst.}) + S. \tag{17}$$

Thus, it appears that the pyro-thiophosphate unit can only be preserved in the glassy state in this system. Also the crystalline phases $Ag_2P_2S_6$ and $Li_2P_2S_6$ form dimers of edge-sharing $PS_{4/2}$ tetrahedra, which do not appear to be represented in the glassy state. The essential results are summarized in Fig. 52. In the glasses, the chemical shift trends observed support the view that the structural environments present here resemble those present in phosphate glasses. Note, however, that in view of the limited chemical shift difference (ca. 8 ppm) and the relatively broad lines the respective centerbands are actually not resolved from each other and the main distinction is afforded by the spinning sideband intensity patterns. Another caveat is that in thiophosphates the shift

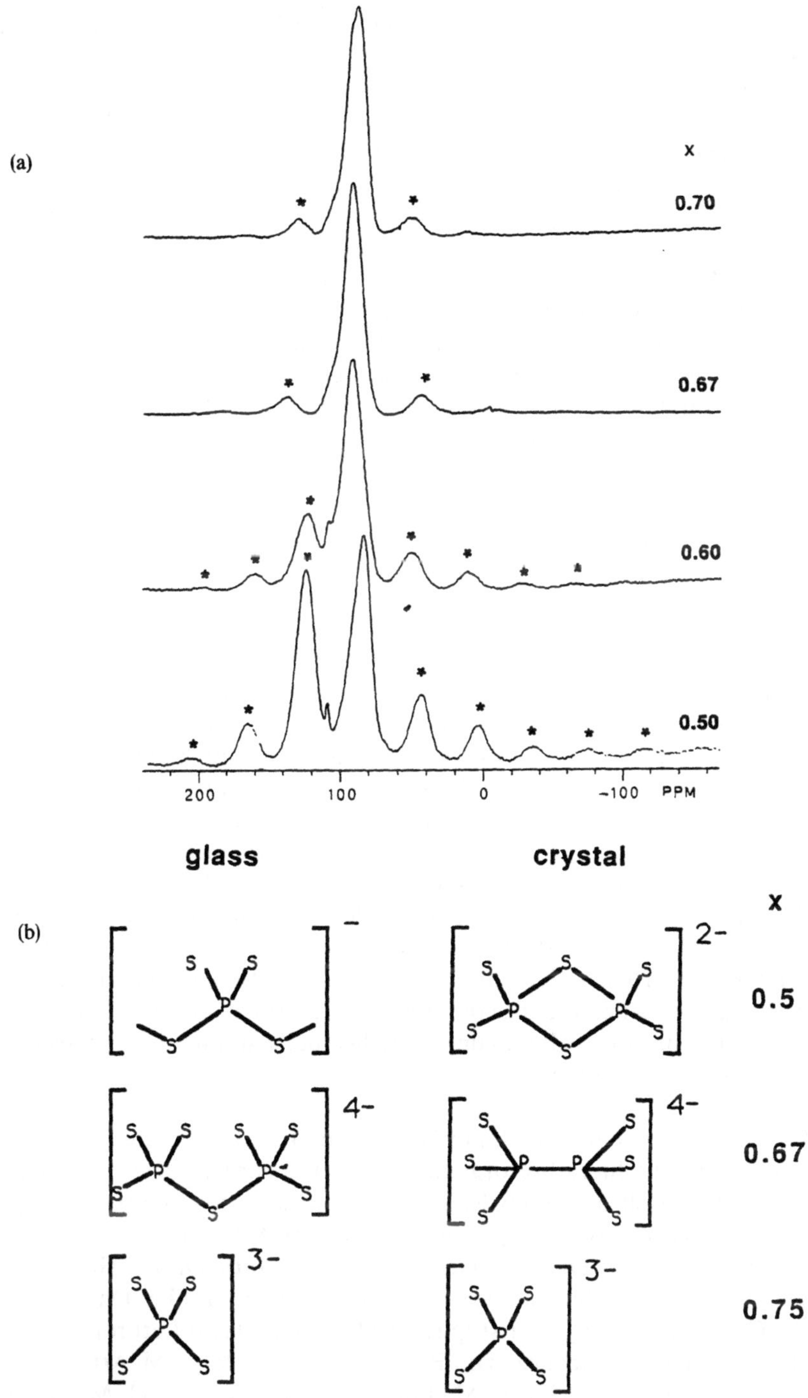

Fig. 52a, b. ^{31}P MAS-NMR spectra **a** and structural interpretation **b** for glasses in the system $(Li_2S)_x(P_2S_5)_{1-x}$. Spinning sidebands are indicated by *asterisks*

discrimination of the various $P^{(n)}$ groups is much less straightforward than in phosphates.

The effect of second network former constituents such as SiS_2, B_2S_3, and Al_2S_3 on the structure of $(Li_2S)_x$–$(P_2S_5)_{1-x}$ has been extensively investigated. The presence of Al_2S_3 favors to some extent the disproportionation reaction (16) in the glassy state and further appears to promote the speciation equilibrium:

$$Al^{(3)} + P^{(1)} \rightarrow Al^{(4)} + P^{(0)}. \tag{18}$$

This scheme is consistent with the well-known tendency of aluminum to avoid nonbridging oxygen species in aluminosilicate glasses. Similarly, in the mixed network former systems with SiS_2 and B_2S_3 phosphorus tends to maximize its number of non-bridging chalcogen atoms (i.e. minimize the number n in the $P^{(n)}$ notation). This phenomenon compares well with the "alkali metal scavenging" effect of phosphorus in the alkali borophosphate and phosphosilicate glasses (see Ref. [4] for recent work in this area).

8 Conclusions

Since the last comprehensive review in this area was written three years ago [4], solid state NMR applications to glasses have been burgeoning, both in quantity and sophistication. Besides covering the incremental progress achieved by virtue of more systematic exploration of individual systems, this review has concentrated on a few landmark articles that have taken this field to new frontiers. Application of increasingly powerful NMR techniques to glasses, the dramatically improved understanding of the chemical shift on a theoretical basis, and the substantial progress made in modeling and simulating the energetics of the glassy state, have all contributed to a much more sophisticated understanding of glass structure than previously thought possible. The vital growth of glass science and the continuous development of new NMR techniques suggests that this area will remain an intellectually stimulating research subject for several years to come.

Acknowledgments. I am grateful to Dr. Deanna Franke for valuable discussions and help with the literature organization at the earlier stages of this article's preparation. I wish to acknowledge the contributions of the past and present members in my research group, Michelle Bankert, Kesha Banks, Mark Beckert, Gary Binas, Nandini Das, Dr. Delphine Davis, Mark Drake, Hans Erickson, Robert Flesher, Dr. Regina Francisco, Dr. Deanna Franke, David Gaughen, Becky Gee, Michael Hay, Christopher Hudalla, Dr. David Lathrop, Tim Luong, Carri Lyda, Ester Martinez, Dr. Robert Maxwell, John McArthur, William McNamara, Dr. Kelly Moran, Michael Morgen, Brian Nadel, Mark Raffetto, Claudia Renteria, David Schmidt, Robert Shibao, James Smith, Melissa St.

Rose, Thomas Tepe, Michael Tullius, Maria Vasquez-Valdez, Juan-Carlos Wallace, Emily Wothe, Yin, Xia, and Dr. Zhengming Zhang. Our research is currently supported by the National Science Foundation, grant DMR 92-21197 and by the UCSB Academic Senate. This support is most gratefully acknowledged.

9 References

1. Zallen R (1986) The physics of amorphous solids. John Wiley, New York
2. Feltz A (1993) Amorphous inorganic materials and glasses. Verlag Chemie, Weinheim
3. Elliott SR (1991) Nature 354: 445
4. Eckert H (1992) Prog. NMR Spectroscopy 24: 159
5. Slichter CP (1989) Principles of magnetic resonance, Springer, Berlin
6. Abragam A (1961) The principles of nuclear magnetism. Clarendon, Oxford
7. Gerstein BC, Dybowski CR (1985) Transient techniques in NMR of solids. Academic, New York
8. Fyfe CA (1984) Solid state NMR for chemists. CRC Press, Guelph, Ontario
9. Mehring M (1983) Principles of high resolution NMR in solids. Springer, Berlin Heidelberg New York
10. Spiess HW (1978) In: Diehl P, Fluck E, Kosfeld R (eds) NMR-Basic Principles and Progress, Vol. 15. Springer Berlin Heidelberg New York, p 55
11. Power WP, Wasylishen RE, Mooibroek S, Pettit, BA, Danchura, W (1990) J Phys Chem 94: 591
12. Cheng, JT, Edwards JC, Ellis P (1990) J Phys Chem 94: 553
13. Kunwar AC, Turner GL, Oldfield E (1986) J Magn Reson 69: 124
14. Rance M, Bird A (1983) J Magn Reson 52: 221

15a. Andrew ER, Bradbury A, Eades RG (1959) Nature 183: 1802

15b. Lowe IJ (1959) Phys Rev Lett 2: 285

16. Samoson A, Kundla E, Lippmaa E (1982) J Magn Reson 49: 350

17a. Mueller KT, Sun BQ, Chingas CG, Zwanziger JW, Terao T, Pines A (1990) J Magn Reson 86: 470

17b. Llor A, Virlet J (1988) Chem Phys Lett 152: 248

17c. Baltisberger JH, Gann SL, Wooten EW, Chang EH, Mueller KT, Pines A (1992) J Am Chem Soc 114: 7489

18. Samoson A Chem Phys Lett (1985) 119: 29
19. Samoson A, Lippmaa E, Pines A (1988) Mol Phys 65: 1013
20. For a comprehensive review of isotropic chemical shift correlation spectroscopy in liquids see: Kessler H, Gehrke M, and Griesinger C (1988) Angew Chem Int Ed Engl 27: 490
21. see for example: Bjorholm T (1988) Chem Phys Lett 143: 259; Fyfe CA, Gies H, Feng Y, Kokotailo GT (1989) Nature 341: 223
22. Knight CTG, Kirkpatrick RJ, Oldfield E (1990) J Noncryst Solids 116: 140
23. Farnan I, Stebbins JF (1990) J Noncryst Solids 124: 207
24. Bjorholm T, Jakobsen H (1991) J Am Chem Soc 113: 27
25. Farnan I, Grandinetti PJ, Baltisberger JH, Stebbins JF, Werner U, Eastman MA, Pines A (1992) Nature 358: 31
26. Frydman L, Chingas GC, Lee YK, Grandinetti PJ, Eastman MA, Barrall GA, Pines A (1992) J Chem Phys 97: 4800
27. Zwanziger JW, Olsen KK, Tagg SL (1993) Phys Rev B 47: 14618

28a. Raleigh DP, Creuzet F, Das Gupta SK, Levitt MH, Griffin RG (1989) J Am Chem Soc 111: 4502

28b. Raleigh DP, Levitt MH, Griffin RG (1988) Chem Phys Lett 146: 71

29a. Levitt MH, Oas TG, Griffin RG (1988) Isr J Chem 28: 271

29b. Oas TG, Levitt MH, Griffin RG (1988) J Chem Phys 89: 692

30a. Gullion T, Schaefer JS (1989) J Magn Reson 81: 196

30b. Gullion T, Schaefer JS (1992) Chem Phys Lett 194: 423

31. Tycko R, Dabbagh G (1991) J Am Chem Soc 113: 9444
32a. Pines A, Gibby MG, Waugh JS (1973) J Chem Phys 59: 569
32b. Hartmann SR, Hahn EL (1962) Phys Rev 128: 2042
33. Hing AW, Vega S, Schaefer JS (1992) J Magn Reson 96: 205
34. Caravatti P, Deli JA, Bodenhausen G, Ernst RR (1982) J Am Chem Soc 104: 5506
35. Moran LB, Berkowitz JK, Yesinowski JP (1992) Phys Rev B 45: 5347
36a. Franke D, Hudalla C, Eckert H (1992) Solid State Nucl Magn Reson 1: 33
36b. Franke D, Maxwell R, Hudalla C, Eckert H (1992) J Phys Chem 96: 7506
37a. Crosby RC, Reese RL, Haw JF (1988) J Am Chem Soc 110: 8550
37b. Fyfe CA, Grondey H, Mueller KT, Wong-Moon KC, Markus T (1992) J Am Chem Soc 114: 5876
37c. Franke D, Hudalla C, Eckert H (1992) Solid State Nucl Magn Reson 1: 297
37d. Franke D, Maxwell R, Eckert H (1992) J Phys Chem 96: 7506
37e. Sebald A, Merwin LH, Schaller T, Knöller W (1992) J Magn Reson 96: 159
37f. Klein Douwel CH, Maas WEJR, Veeman GS, Werumeus Buning GH, Vankan JM (1990) Macromolecules 23: 406
38. Lippmaa E, Samoson A, Mägi M (1986) J Am Chem Soc 108: 1730
39. Kunath G, Losso P, Steuernagel S, Schneider H, Jäger C (1992) Solid State NMR 1: 261
40a. Engelsberg M, Norberg RE (1972) Phys Rev B 5: 3395
40b. Boden N, Gibb M, Levine YK, Mortimer M (1974) J Magn Reson 16: 471
41. Reimer JA, Duncan TM (1983) Phys Rev B 27: 4895
42a. Duncan TM, Douglass DC, Csencsits R, Walker KL (1986) J Appl Phys 60: 130
42b. Duncan TM, Douglass DC, Walker KL, Csencsitis R (1985) J Appl Phys 58: 197
42c. Lathrop D, Eckert H (1989) J Am Chem Soc 114: 3536
43a. Kaplan DE, Hahn EL (1958) J Phys Radium 19: 821
43b. Shore SE, Ansermet JP, Slichter CP, Sinfelt JH (1987) Phys Rev Lett 58: 953
43c. Wang PK, Slichter CP, Sinfelt JH (1984) Phys Rev Lett 53: 82
44. Boyce JB, Ready SE (1988) Phys Rev B38: 11008
45. Lathrop D, Eckert H (1990) J Am Chem Soc 112: 9017
46. Van Vleck JH (1948) Phys Rev 74: 1168
47. Franke D, Hudalla C, Eckert H (1992) Solid State Nucl Magn Reson 1: 73
48a. Baum J, Gleason KK, Pines A, Garroway AN, Reimer JA (1986) Phys Rev Lett 56: 1377
48b. Gleason KK, Petrich MA, Reimer JA (1987) Phys Rev B 36: 3259
48c. Baum J, Pines A (1986) J Am Chem Soc 108: 7447
49a. Scruggs BE, Gleason KK (1992) Chem Phys 166: 367
49b. Gleason KK (1993) Concepts Magn Reson 5: 199
49c. Cho G, Yesinowski JP (1993) Chem Phys Lett 205: 1
50. Gaskell PH, Eckesley MC, Barnes AC, Chieux P (1991) Nature (London), 350: 675
51. Gurman SJ (1990) J Noncryst Solids 125: 151
52a. Grimmer AR (1985) Chem Phys Lett 119: 416
52b. Lippmaa E, Mägi M, Samoson A, Engelhardt G, Grimmer AR (1980) J Am Chem Soc 102: 4889
52c. Mägi M, Lippmaa E, Samoson A, Engelhardt G, Grimmer AR (1984) J Phys Chem 88: 1518
53a. Stebbins JF (1987) Nature 330: 465
53b. Stebbins JF (1988) J Noncryst Solids 106: 359
53c. Emerson JF, Stallworth PE, Bray PJ (1989) J Noncryst Solids 113: 253
54a. Dupree R, Holland D, Williams DS (1984) J Noncryst Solids 68: 399
54b. Grimmer AR, Mägi M, Hähnert M, Stade H, Samoson A, Wieker W, Lippmaa E (1984) Phys Chem Glasses 25: 105
54c. Schramm CH, deJong BHWS, Parziale VE (1984) J Am Chem Soc 106: 4396
54d. Hater W, Müller-Warmuth W, Meier M, Frischat GH (1989) J Noncryst Solids 113: 210
54e. Hater W, Müller-Warmuth W, Frischat GH (1989) Glastech Ber 62: 328
54f. Dupree R, Holland D, Mortuza MG (1990) J Noncryst Solids 116: 148
54g. Dupree R, Holland D, Williams DS (1986) J Noncryst Solids 81: 185
54h. Schneider E, Stebbins JF, Pines A (1987) J Noncryst Solids 89: 371
55. Maekawa H, Maekawa T, Kawamura K, Yokokawa T (1991) J Noncryst Solids 127: 53
56. Grimmer AR, Müller D (1986) Monatsh Chem 117: 799
57. Uchino T, Sakka T, Ogata Y, Iwasaki M (1992) J Phys Chem 96: 2455
58. Murdoch JB, Stebbins JF, Carmichael ISE (1985) Am Miner 70: 332
59a. Pettifer RF, Dupree R, Farnan I, Sternberg U (1988) J Noncryst Solids 106: 408

59b. Dupree R, Kohn SC, Henderson CMB, Bell AMT (1993) In: Tossell JA (ed) Nuclear magnetic shiedings and molecular structure. Kluwer, Dordrecht, p 421
60. Sternberg U (1993) In: Tossell JA (ed) Nuclear magnetic shieldings and molecular structure. Kluwer, Dordrecht, p 435
61. Tossell JA, Lazzeretti P (1988) Phys Chem Miner 15: 564
62. Mueller KT, Pines A (1990) J Magn Reson 86: 470
63. Geissberger AE, Bray PJ (1983) J Noncryst Solids 54: 121
64. Mozzi RL, Warren BE (1969) J Appl Cryst 2: 164
65. Melman H, Garofalini SH (1991) J Noncryst Solids 134: 107
66. Jäger C, Dupree R, Kohn SC, Mortuza MG (1993) J Noncryst Solids 155: 95
67. Farnan I, Stebbins JF (1990) J Am Chem Soc 112: 32
68. Farnan I, Stebbins JF (1990) J Noncryst Solids 124: 207
69. Stebbins JF (1991) Nature 351: 638
70. Xue X, Stebbins JF, Kanzaki M, McMillan PF, Poe B (1991) Am Miner 76: 8
71. Angell CA, Scamehorn CA, Phifer CC, Kadiyala RR, Cheseman PA (1988) Phys Chem Miner 15: 221
72. Dupree R, Holland D, Mortuza MG, Collins JA, Lockyer MWG (1989) J Noncryst Solids 112: 111
73. Krogh-Moe J (1969) J Noncryst Solids 1: 269, and references therein.
74a. Bray PJ (1974) Proc 10th Int Congr Glass, Kyoto Ceramic Society of Japan, p 13
74b. Müller-Warmuth W, Eckert H (1982) Phys Rep 88: 91
74c. Bray PJ, Geissberger AE, Bucholtz F, Harris IA (1982) J Noncryst Solids 52: 45
74d. Bray PJ (1985) J Noncryst Solids 73: 19
74e. Bray PJ (1987) J Noncryst Solids 95/96: 45
74f. Bray PJ, Gravina SJ, Hintenlang DH, Mulkern RV (1988) Magn Reson Rev 13: 263
75. Jellison Jr. GE, Panek LW, Bray PJ, Rouse Jr GB, Risen WM (1977) J Chem Phys 66: 802
76. Gravina SJ, Bray PJ (1990) J Magn Reson 89: 515
77. Lee D, Gravina SJ, Bray PJ (1990) Z Naturforsch 45a: 268
78. Gravina SJ, Bray PJ, Petersen DL (1990) J Noncryst Solids 123: 165
79. Bray PJ, Lee D, Mao DG, Petersen GL, Feller SA, Bain DL, Feil DA, Pandikuthira P, Nijhawan S (1992) Z Naturforsch 47a: 30
80. Bray PJ, Emerson JF, Lee D, Feller SA, Bain DL, Feil DA (1991) J Noncryst Solids 129: 240
81. Mao D, Bray PJ (1992) J Noncryst Solids 144: 217
82. Connor C, Chang J, Pines A (1990) J Chem Phys 93: 7639
83. Youngman RE, Zwanziger JW (1994) J Noncryst Solids, in press.
84. Tossell JA, Lazzeretti P (1988) J Noncryst Solids 99: 267
85. Tossell JA (1990) J Noncryst Solids 120: 13
86. Zhong J, Bray PJ (1989) J Noncryst Solids 111: 67
87. Van Wüllen L, W. Müller-Warmuth W (1993) Solid State NMR 2: 279
88. Zhonghong J, Yongxing T (1992) J Noncryst Solids 146: 57
89. Yongxing T, Zhonghong J, Xiuyu S (1989) J Noncryst Solids 112: 131
90. Dell WJ, Bray PJ, Xiao SZ (1983) J Noncryst Solids 58: 1
91. Brow RK, Kirkpatrick RJ, Turner GL (1990) J Noncryst Solids 116: 39
92. Hayashi S, Hayamizu K (1989) J Solid State Chem 80: 195
93. Losso P, Schnabel B, Jäger C, Sternberg U, Stachel D, Smith DO (1992) J Noncryst Solids 143: 265
94. Brow RK (personal communication) to be published
95. Grimmer AR (1978) Spectrochim Acta 34A: 941
96a. Villa M, Carduner KR, Chiodelli G (1987) J Solid State Chem 69: 19
96b. Villa M, Carduner KR, Chiodelli G (1987) Phys Chem Glasses 28: 131
96c. Villa M, Scagliotti M, Chiodelli G (1987) J Noncryst Solids 94: 101
97. Brow RK, Kirkpatrick RJ, Turner GL (1990) J Am Ceram Soc 73: 2293
98. Brow RK, Kirkpatrick RJ, Turner GL (1993) J Am Ceram Soc 76: 919
99. Sato RK, Kirkpatrick RJ, Brow RK (1992) J Noncryst Solids 143: 257
100. Fletcher JP, Risbud SH, Kirkpatrick RJ (1990) J Mater Res 5: 835
101. Anma M, Yano T, Yasumori A, Kawazoe H, Yamane M, Yamanaka H, Katada M (1991) J Noncryst Solids 135: 79
102. Brow RK, Osborne ZA, Kirkpatrick RJ (1992) J Mater Res 7: 1892
103. Engelhardt G, Nofz M, Forkel K, Wihsmann FG, Mägi M, Samoson A, Lippmaa E (1985) Phys Chem Glasses 26: 157

104. Oestrike R, Yang WH, Kirkpatrick RJ, Hervig RL, Navrotsky A, Montez B (1987) Geochim Cosmochim Acta 51: 2199
105. Kirkpatrick RJ, Oestrike R, Weiss Jr CA, Smith KA, Oldfield E (1986) Am Miner 71: 705
106. Oestrike R, Kirkpatrick RJ (1988) Am Miner 73: 534
107. Merzbacher CI, Sherriff BL, Hartmann JS, White WB (1990) J Noncryst Solids 124: 194
108. Sato RK, McMillan PF, Dennison P, Dupree R (1991) Phys Chem Glasses 32: 149
109. Loewenstein W (1954) Am Miner 39: 92
110. McMillan PF, Kirkpatrick RJ (1992) Am Mineral 77: 898
111. Maekawa H, Maekawa T, Kawamura K, Yokokawa T (1991) J Phys Chem 95: 6822
112. Landron C, Cote B, Massiot D, Coutures JP, Flank AM (1992) Phys Status Solidi (b) 171: 9
113. Cote B, Massiot D, Poe P, McMillan PF, Taulelle F, Coutures JP (1992) J Physique IV C2: 223
114. Poe BT, McMillan PF, Cote B, Massiot D, Coutures JP (1993) Science 259: 786
115. Higby PL, Ginther RJ, Aggarwal ID, Friebele EJ (1990) J Noncryst Solids 126: 209
116. Merzbacher CI, McGrath KJ, Higby PL (1991) J Noncryst Solids 136: 249
117. Lacy ED (1963) Phys Chem Glasses 4: 234
118a. Jantzen CM, Herman HJ (1979) Am Ceram Soc 62: 212
118b. Risbud SH, Pask JA (1979) J Am Ceram Soc 62: 214
119. Risbud SH, Kirkpatrick RJ, Tagliavore AP, Montez B (1987) J Am Ceram Soc 70: C10
120. Yasumori A, Iwasaki M, Kawazoe H, Yamane M, Nakamura Y (1990) Phys Chem Glasses 31: 1
121. Sato RK, McMillan PF, Dennison P, Dupree R (1991) J Phys Chem 95: 4483
122. Prabhakar S, Rao KJ, Rao CNR (1992) Eur J Solid State Inorg Chem 29: 95
123. Poe BT, McMillan PF, Angell CA, Sato RK (1992) Chem Geol 96: 333
124. Poe BT, McMillan PF, Cote B, Massiot D, Coutures JP (1992) J Phys Chem 96: 8220
125. Farnan I, Dupree R, Jeong Y, Thompson GE, Wood GC, Forty AJ (1989) Thin Solid Films 173: 209
126. Dupree R, Farnan I, Forty AJ, El-Mashri S, Bottyan L (1985) J Physique C8, 46: 113
127. Wood TE, Siedle AR, Hill JR, Skarjune RP, Goodbrake CJ (1990) Mater Res Soc Symp Proc 180: 97
128. Nazar LF, Fu G, Bain AD (1992) J Chem Soc Chem Commun 251
129. Kohn SC, Dupree R, Mortuza MG, Henderson CMB (1991) Am Mineral 76: 309
130. Schaller T, Dingwell DB, Keppler H, Knöller W, Merwin L, Sebald A (1992) Geochim Cosmochim Acta 56: 701
131. Tossell JA (1993) Am Mineral 78: 16
132. Hyatt MJ, Day DE (1987) J Am Ceram Soc 70: C283
133. Shelby JE, Kohli JT (1990) J Am Ceram Soc 73: 39
134. Kohli JT, Shelby JE, Frye JS (1992) Phys Chem Glasses 33: 73
135a. Bishop SG, Bray PJ (1966) Phys Chem Glasses 7: 73
135b. Gresch R, Müller-Warmuth W, Dutz H (1976) J Noncryst Solids 21: 31
136a. Dupree R, Holland D, Williams DS (1985) Phys Chem Glasses 26: 50
136b. Oestrike R, Navrotsky A, Turner GL, Montez B, Kirkpatrick RJ (1987) Am Mineral 72: 788
136c. Hähnert M, Hallas E (1987) Rev Chim Miner 24: 221
136d. Hähnert M, Hallas E (1986) Z Chem 26: 144
136e. Hallas E, Gerth K, Hähnert M (1987) Z Chem 27: 270
137. Bunker BC, Kirkpatrick RJ, Brow RK, Turner GL, Nelson C (1991) J Am Ceram Soc 74: 1430
138. Bunker BC, Kirkpatrick RJ, Brow RK (1991) J Am Ceram Soc 74: 1425
139. Müller D, Berger G, Grunze I, Ladwig G, Hallas E, Haubenreisser U (1983) Phys Chem Glasses 24: 37
140. Brow RK, Kirkpatrick RJ, Turner GL (1990) J Am Ceram Soc 69: C222
141a. Kunath G, Losso P, Steuernagel S, Schneider H, Jäger C (1992) Solid State NMR 1: 261
141b. Jäger C (1992) J Magn Reson 99: 353
142. Jäger C, Müller-Warmuth W, Mundus L, Van Wüllen L (1992) J Noncryst Solids 149: 209
143. Liu Z, Taylor PC (1989) Solid State Commun 70: 81
144. Phillips JC (1979) J Noncryst Solids 34: 153
145. Thorpe MF (1985) J Noncryst Solids 76: 109
146. Tatsumisago T, Halfpap BL, Green JL, Lindsay SM, Angell CA (1990) Phys Rev Lett 64: 1549
147a. Tenhover M, Hazle MA, Grasselli RK (1984) Phys Rev B 29: 6732
147b. Tenhover M, Henderson RS, Lukco D, Hazle MA, Grasselli RK (1984) Solid State Commun 51: 455
147c. Griffiths JE, Malyi M, Espinosa GP, Remeika JP (1984) Phys. Rev B 30: 6978

148. Krebs B (1983) Angew Chem 95: 113
149. Tenhover M, Boyer RD, Henderson RS, Hammond TE, Shreve GA (1988) Solid State Commun 65: 1517
150. Moran KL, Shibao RK, Eckert H (1990) Hyperfine Int 62: 55
151. Shibao RK, Hay M, Srdanov V, Eckert H (1994) Chem Mater 6: 306
152. Tenhover M, Henderson RS, Hazle MA, Lukco D, Grasselli RK (1987) In: Cocke DL, Clearfield A (eds) Design of New Materials, Plenum Publishing Corp p 329
153. Pradel A, Michel-Lledos V, Ribes M, Eckert H (1993) Chem Mater 5: 377
154a. Borisova ZU (1981) Glassy semiconductors. Plenum, New York, p 70
154b. Borisova ZU, Kasatkin BE, Kim EI (1973) Izv Akad Nauk SSSR Neorg Mater 9: 822
155. Blachnik R, Hoppe A (1979) J Noncryst Solids 34: 191
156a. Monteil Y, Vincent H (1975) J Inorg Nucl Chem 37: 2053
156b. Monteil Y, Vincent H (1977) Z Anorg Allg Chem 428: 259
156c. Monteil Y, Vincent H (1974) Can J Chem 52: 2190
157. Heyder F, Linke D (1973) Z Chem 13: 480
158. Kim EI, Chernov AP, Dembovskii SA, Borisova ZU (1976) Izv Akad Nauk SSSR Neorg Mater 12: 1021
159. Price DL, Misawa M, Susman S, Morrison TI, Shenoy GK, Grimsditch M (1984) J Noncryst Solids 66: 443
160. Arai M, Johnson RW, Price DL, Susman S, Gay M, Enderby JE (1986) J Noncryst Solids 83: 80
161. Kumar A, Malhotra LK, Chopra KL (1987) J Noncryst Solids 92: 51
162. Lathrop D, Eckert H (1988) J Noncryst Solids 106: 417
163. Lathrop D, Eckert H (1989) J Phys Chem 93: 7895
164. Lathrop D, Eckert (1989) J Am Chem Soc 111: 3536
165. Eckert H, Franke D, Lathrop D, Maxwell R, Tullius M (1990) Mat Res Soc Symp Proc 172: 193
166. Lathrop D, Eckert H (1991) Phys Rev B 43: 7279
167. Lathrop D, Eckert H (1990) J Am Chem Soc 112: 9017
168. Maxwell R, Eckert H (1993) J Am Chem Soc 115: 4747
169. Maxwell R, Eckert H (1994) J Am Chem Soc 116: 682
170. Lathrop D, Tullius M, Tepe T, Eckert H (1991) J Noncryst Solids 128: 208
171. Lathrop D, Eckert H (1993) J Noncryst Solids 160: 111
172. Lyda C, Leone J, Tepe T, Lathrop D, Eckert H to be published.
173. Demarcq MC (1987) Phosphorus Sulfur 33: 127
174. Demarcq MC (1990) J Phys Chem 94: 7330
175. Tullius M, Lathrop D, Eckert H (1990) J Phys Chem 94: 2145
176a. Eckert H, Liang CS, Stucky GD (1989) J Phys Chem 93: 452
176b. Harris RK, Wilkes PJ, Wood PT, Woollins JD (1989) J Chem Soc Dalton Trans 809
177. Koudelka L, Pisarek M, Blinov LN, Gutenev MS (1991) J Noncryst Solids 134: 86
178. Bjorholm T (1988) Chem Phys Lett 143: 259
179. Vaipolin AA, Osmanov EO, Rud VV (1966) Sov Phys Solid State 7: 1833
180. Goryunova NA, Kuzmenko GS, Osmanov EO (1971) Mater Sci Eng 7: 54
181a. Polk DE (1971) J Noncryst Solids 5: 365
181b. Polk DE, Boudreux DS (1973) Phys Rev Lett 31: 92
181c. Steinhardt P, Alben R, Weaire D (1974) J Noncryst Solids 15: 199
182a. Wooten F, Weiner K, Weaire D (1985) Phys Rev Lett 54: 1392
182b. Wooten F, Weaire D (1984) J Noncryst Solids 64: 325
182c. Wejchert J, Weaire D, Wooten F (1990) J Noncryst Solids 122: 241
183a. Greenbaum SG, Marino RA, Adamic KJ, Case C (1985) J Noncryst Solids 77/78: 1285
183b. Greenbaum SG, Treacy DJ, Shanabrook BV, Comas J, Bishop SG (1984) J Noncryst Solids 66: 133
184. Franke D, Maxwell R, Lathrop D, Eckert H (1991) J Am Chem Soc 13: 4822
185. Franke D, Maxwell R, Lathrop D, Banks K, Eckert H (1992) Phys Rev B 46: 8109
186. Franke D, Eckert H (1991) J Phys Chem 95: 331
187. Thornburg D (1977) Thin Solid Films 45: 95
188. Hong KS, Berta Y, Speyer RF (1990) J Am Ceram Soc 73: 1351
189. Mahadevan S, Giridhar A (1991) J Noncryst Solids 128: 8
190a. Malugani JP, Robert G (1980) Solid State Ionics 1: 519
190b. Pradel A, Ribes M (1986) Solid State Ionics 18/19: 351
190c. Zhang Z, Kennedy JH (1990) Solid State Ionics 38: 217

190d. Kennedy JH, Zhang Z, Eckert H (1990) J Noncryst Solids 123: 328
191. Hürter HU, Krebs B, Eckert H, Müller-Warmuth W (1985) Inorg Chem 24: 1288
192. Eckert H, Müller-Warmuth W, Hamann W, Krebs B (1984) J Noncryst Solids 65: 53
193. Hintenlang DE, Bray PJ (1985) J Noncryst Solids 69: 243
194. Zahir M, Villeneuve G, Olazcuaga R (1985) Rev Chim Miner 22: 297
195. Zhang Z, Kennedy JH, Thompson J, Anderson S, Lathrop D, Eckert H (1989) Appl Phys A 49: 41
196. Suh KS, Hojjaji A, Villeneuve G, Menetrier M, Levasseur A (1991) J Noncryst Solids 128: 13
197a. Zum Hebel P, Krebs B, Grüne M, Müller-Warmuth W (1990) Solid State Ionics 43: 133
197b. Hiltman R, Krebs B (1993) Z Anorg Allg Chem 619: 293
198. Eckert H, Zhang Z, Kennedy JH (1989) J Noncryst Solids 107: 271
199. Eckert H, Kennedy JH, Pradel A, Ribes M (1989) J Noncryst Solids 113: 287
200. Pradel A, Ribes M, Eckert H (to be published)
201. Martin SW, Sills JA (1991) J Noncryst Solids 135: 171
202. Pradel A, Michel-Lledos V, Ribes M, Eckert H (1992) Solid State Ionics 53–56: 1187
203. Eckert H, Zhang Z, Kennedy JH (1990) Chem Mater 2: 273
204. Zhang Z, Kennedy JH, Eckert H (1992) J Am Chem Soc 114: 5775
205. Eckert H, Zhang Z, Kennedy JH (1989) Mater Res Soc Symp Proc 135: 259
206. Kennedy JH, Schaupp CS, Eckert H, Ribes M (1991) Solid State Ionics 45: 21

Author Index Volumes 21-33

MIX
Papier aus verantwortungsvollen Quellen
Paper from responsible sources
FSC® C105338

If you have any concerns about our products,
you can contact us on
ProductSafety@springernature.com

In case Publisher is established outside the EU,
the EU authorized representative is:
Springer Nature Customer Service Center GmbH
Europaplatz 3, 69115 Heidelberg, Germany

Printed by Libri Plureos GmbH
in Hamburg, Germany